SAM MALOOF

36 Views of a Master Woodworker

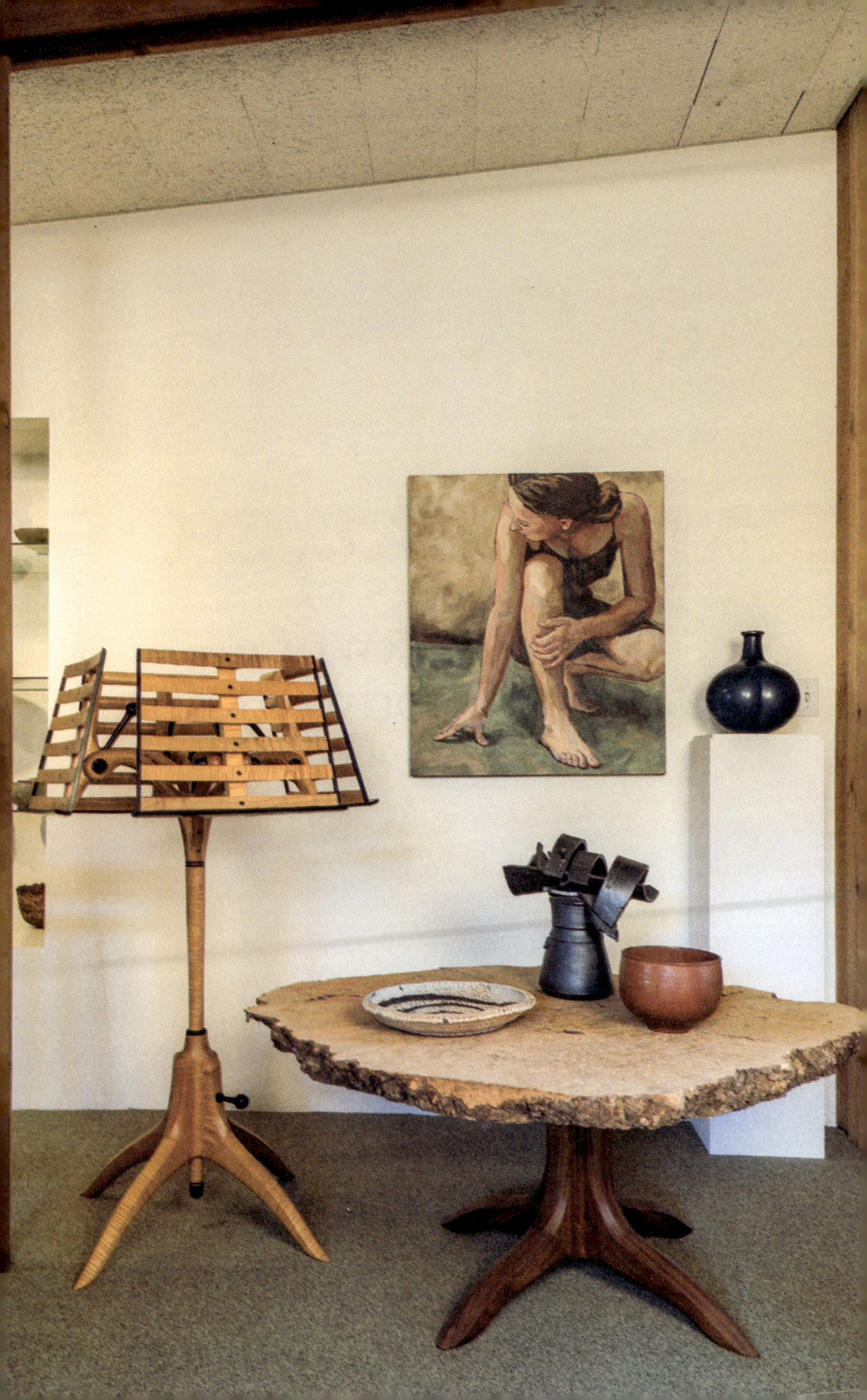

SAM MALOOF

36 Views of a Master Woodworker

Fred Setterberg

Heyday, Berkeley, California
The Sam and Alfreda Maloof Foundation for Arts and Crafts,
Alta Loma, CA

Library of Congress Cataloging-in-Publication Data

Setterberg, Fred, author.
Sam Maloof : 36 views of a master woodworker / by Fred Setterberg.
pages cm
ISBN 978-1-59714-333-2 (hardcover : alk. paper)
1. Maloof, Sam--Anecdotes. 2. Woodworkers--United States--Anecdotes. 3. Cabinetmakers--United States--Anecdotes. I. Title.
NK2439.M28S48 2016
749.092--dc23
2015031650

Cover Photo: Jim McHugh
Book Design: Ashley Ingram

Special thanks to Toni and Tom Bostick for contemporary views of the Maloof Historic Home and Workshop, and to Sam Maloof Woodworker, Inc., for the following images: photos on pages 24, 150, and 216 are courtesy of Schenck and Schenck Photography; photo on page 124 is courtesy of Gene Sasse.

Sam Maloof was published by Heyday and the Sam and Alfreda Maloof Foundation for Arts and Crafts. Orders, inquiries, and correspondence should be addressed to:

Heyday
P.O. Box 9145, Berkeley, CA 94709
(510) 549-3564, Fax (510) 549-1889
www.heydaybooks.com

Printed in Stevens Point, WI, by Worzalla

10 9 8 7 6 5 4 3 2 1

For Malcolm—for more reasons than I can count....

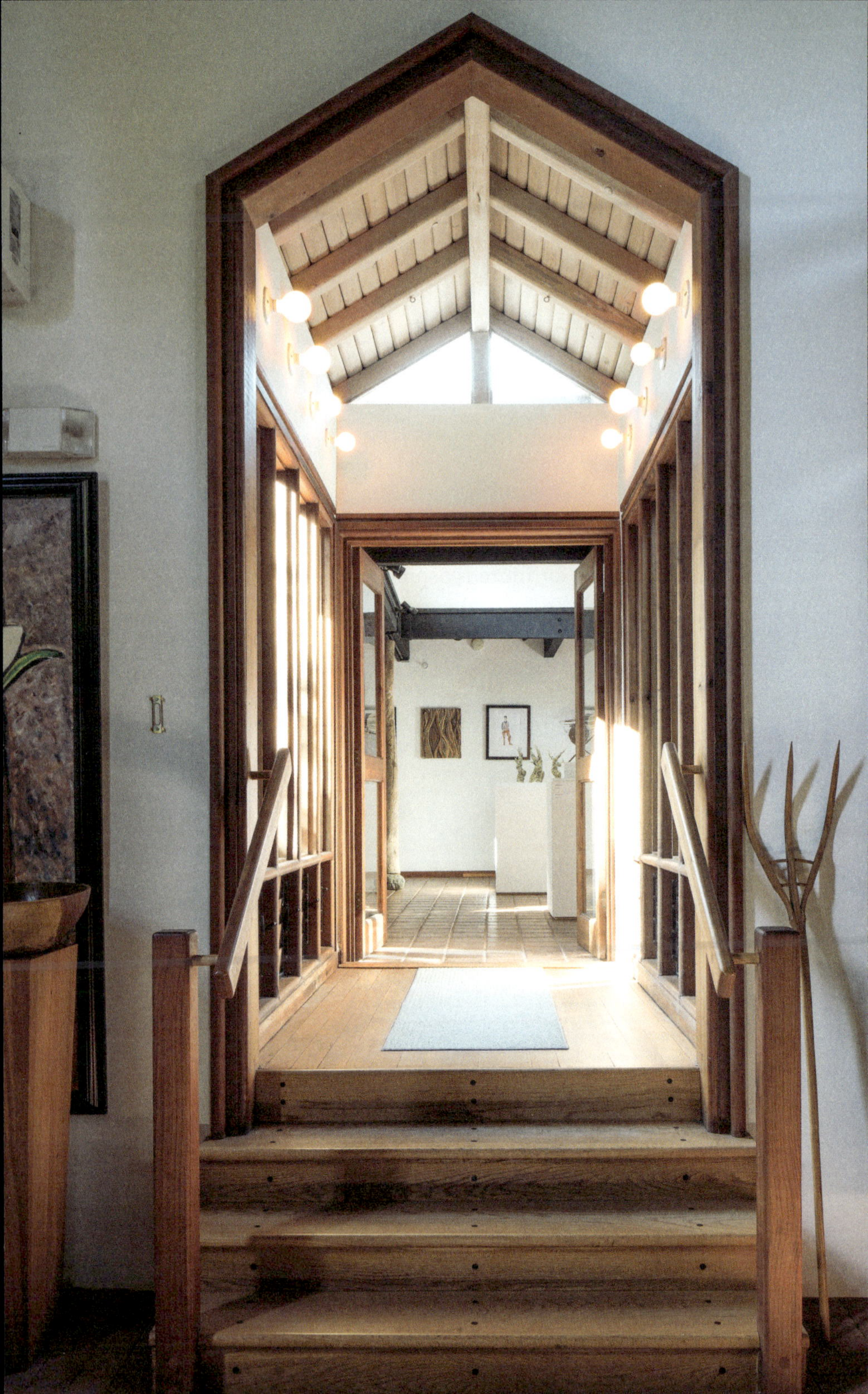

The bird a nest, the spider a web, man friendship.

—William Blake

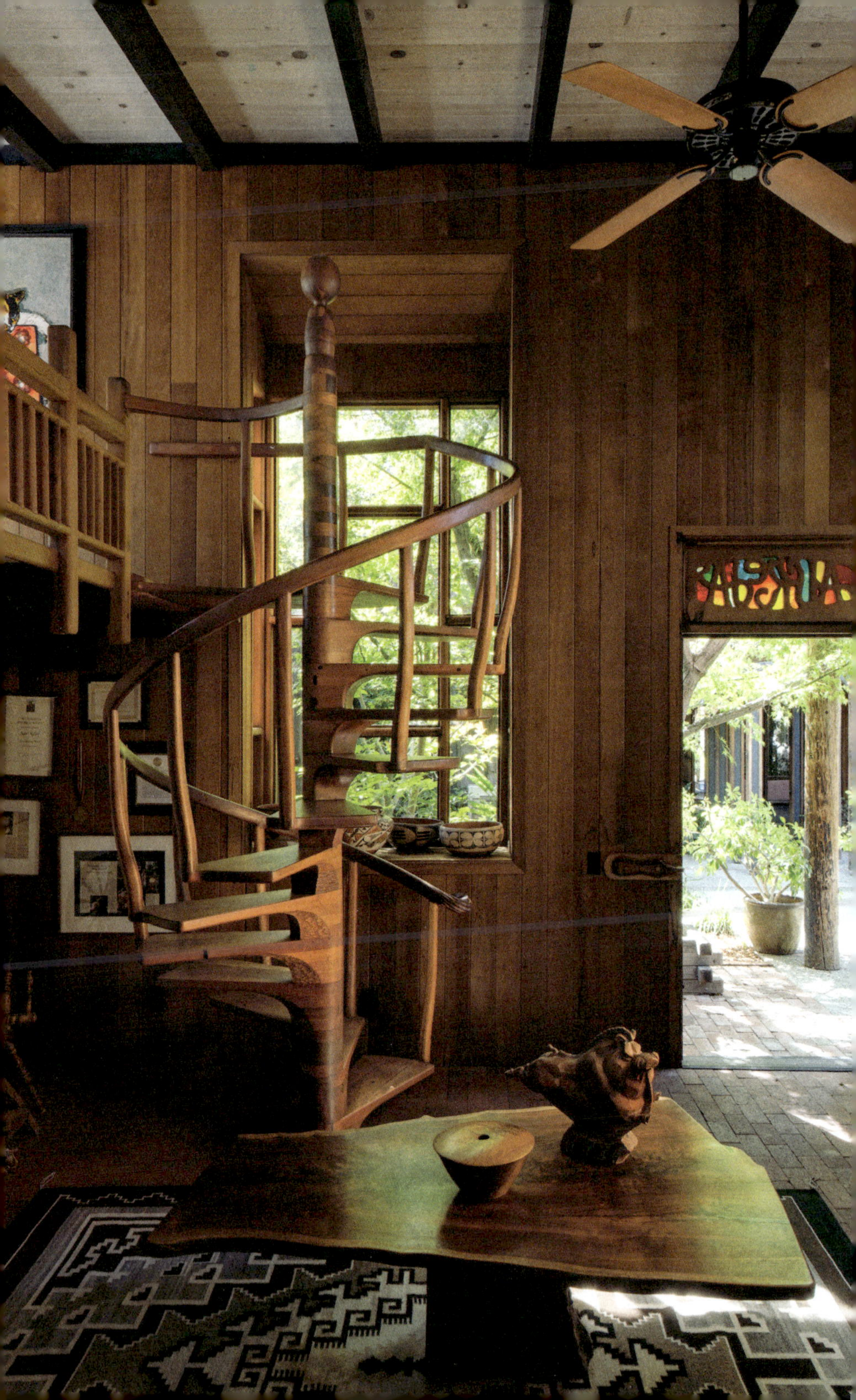

CONTENTS

PREFACE
REMEMBERING SAM

How do you describe the effect and appeal of Sam Maloof's artistry? His furniture radiates warmth; it's alive to the touch, sensuous, inviting—you want to reach out, caress the flats of his tables, lay your hands across the shoulders of his chairs, ease your backside into the implausibly commodious wooden seats scooped from his rockers.

"I'm sure you've heard all this before," acknowledged a friend who used to visit the Maloof home with her parents in the 1950s, roaming through the house as it grew by innovations and board feet, curling herself into the chairs and settees while Sam cut and shaped another table or cradle in the shop.

And it's true. The words deployed most frequently by the owners and admirers of Sam's furniture, and by visitors to his sculpted home, evoke an intense pleasure and admiration for his work, a devotion seldom lavished on furniture or domestic architecture. But while Sam's closest friends and most informed enthusiasts avidly respond to queries about the impact of his art, they also move the conversation in unexpected directions. With immense feeling and sharp recollection, they describe how Sam, the artist and friend, affected them. And oftentimes, how they influenced Sam.

Sam Maloof was one of the great practitioners of midcentury modernism, an artist of far-reaching influence who took pleasure in describing himself simply as a woodworker. In 1985, he became the first craftsperson to receive a MacArthur Foundation "Genius Award," achieving widespread national prestige and eliding any remaining distance in critical perception between craft and art. Starting out in the 1950s, he designed and built some five thousand pieces of furniture for homes, offices, and public institutions, while simultaneously transforming what he described as an eight-hundred-square-foot "dingbat" bungalow into a twenty-two-room residence now listed on the National Register of Historic Places. He sold directly to his clientele, working out of his shop in Alta Loma, California, embracing a style founded in the Arts and Crafts movement, borrowing heavily from Danish modern, and evolving slowly, but resolutely into the unmistakable Maloof style. His work is collected by the Smithsonian, the Metropolitan Museum of Art, the Boston Museum of Fine Arts, the White House Craft Collection, and numerous other museums and private enthusiasts. When Sam died in 2009, he was widely regarded as the most influential woodworking artist of his era. Or in the words of his friend and fellow furniture maker, Jimmy Carter, "the best woodworker that ever lived."

Ironically, Sam's reputation and legacy is secured in part by one of the most painful and disruptive episodes of his long life. In 1990, the California State Legislature approved funds for a freeway extension in San Bernardino County aimed directly at the Maloof home and property. Fortunately, Sam's recognition as an important artist—indeed, a key figure in California midcentury modernism taking root in the Pomona Valley—enabled a team of friends and backers to successfully lobby for the house's relocation some two and a half miles away on property now belonging to the Sam and Alfreda Maloof Foundation for Arts and Crafts.

The group effort involved in rescuing Sam's house and establishing an institutional basis for the preservation of his work and reputation aligns with an essential aspect of his career. For all his imagination and productivity sustained over six decades, the example of Sam Maloof undermines the notion that art is created and nurtured in isolation. We celebrate the protean figure, the artist as individualist, even as remote loner. Far more frequently, the artist's life and creative output is interwoven imperfectly into the fabric of family, neighborhood, far-flung community, alliance and rivalry. Sam composed his art in intimate contact with a huge number of friends and family, in connection with a remarkable array of artists throughout the Pomona Valley and others farther afield.

Who were Sam's direct collaborators? Foremost, Alfreda Ward Maloof, Sam's wife, partner of fifty years, and indispensible ally. (Sam remarried several years following Alfreda's death to a longtime friend, Beverly Wingate.) Certainly, "the boys"—a select cadre of young men who typically began apprenticing in Sam's shop in their late teens and early twenties, with several ripening alongside him into middle age and beyond.

Sam also benefitted from his chance landing in the right place and time. The Pomona Valley arts scene progressed along the same scaffold of calendar years as Sam's career. Millard Sheets, the seminal figure of the Claremont Colleges' midcentury expansion into the visual arts, served as Sam's early employer and mentor. Thanks to Millard, Sam befriended a gallery of peers recently discharged from the military who were using their G.I. benefits to study and kick-start careers in painting, sculpture, and ceramics. Good friends made better artists. Sam and his colleagues mixed, married, experimented, and matured in small-town pastoral seclusion, where they could refine and extend the dictum of the Arts and Crafts movement to embrace, own, and create only the useful and the beautiful—while reframing both terms with the refracting lens of modernism.

Today, Sam continues to attract admiration as the plentiful author of furniture as functional art and the creator of an artist's home that evinces wonder and appreciation from strangers now that he has gone. But his garrulous sociability, his talent for comradeship, also provides us with an unusually rich opportunity to recall his life and career from literally dozens of perspectives.

This book aims to offer a kaleidoscopic portrait of Sam Maloof from the varied, contradictory, intensely felt experience of his family, friends, and associates—an affectionately inconstant view of the master craftsman and artist in his time and place. A man whose spirit and example touched an ever-expanding sphere of friends and colleagues with the same consequential influence as his art.

1.

MAKING ART
LARRY WHITE

One morning in June 1962, the placement director at Chaffey College called Larry White with a job lead. Since enrolling at the community college, Larry had been haunting her office, picking up part-time work whenever possible—an industriousness that eased his entry into the study of ceramics, a field unattached to expectations of predictable income.

"Do you know who Sam Maloof is?" she asked.

He didn't.

Sam, she explained, was a well-known artist living about four miles from the college. He was looking for a helper. Summer only. Larry should give him a call.

What she might have said: *Today I am going to change your life.*

"I grew up in a little box in Ontario," said Larry. "When I walked into Sam's compound and saw the architecture and the ceramics and the gates he had made, I knew I was heading into an entirely different world."

Larry had already been accepted to complete his studies at the San Francisco Art Institute, and he was debating whether to attend. "Then I got the job with Sam and decided just to follow that wherever it went. I met his friends—

other craftsmen, artists, designers. Amazing people. Some I'd heard of, some I hadn't. Sam kept reiterating that my job was only for the summer. He was very apprehensive about continuing to support an employee beyond that. He'd just come back from Iran, where he'd been for six months or so with the State Department, helping local craftspeople create indigenous products for export. When he returned, he was basically broke and way behind in his orders."

At the end of the summer, Sam deemed their relationship a success and suggested they continue, filling in hours around Larry's schedule.

"That led to a seven-year apprenticeship," said Larry. "Sam had then been working for sixteen years by himself. I was his first paid employee."

At least, that's the tale that Larry has often recounted over the past fifty years.

"Recently," he said, "I found out there's a backstory."

Sam, it turns out, had asked his friend, the ceramicist Tony Ivins, if he had any students at Chaffey whom he considered "really good." Sam didn't want just anybody. He required discipline, passion, somebody capable with his hands and possessed of a strong aesthetic—a battery of requirements not easily reduced to an index card in the placement office or extracted from a résumé. Tony recommended Larry. Sam asked the placement director to make the call. "That's how we hooked up," said Larry. Less accident than fate with a bump. Or, as time would prove, the plausible meshing of compatible parts.

"I knew how to work, how to focus. I was familiar with hardwood because my dad was an amateur gunsmith—he did custom stocks. I knew about inlays and exotics. But what Sam was doing was totally unfamiliar to me. Making art furniture."

The key word, *art.*

At twenty-one, Larry was discovering in himself a compulsion to create. He drew, painted, sculpted, threw pots and worked with clay and glazes. The day didn't contain enough hours to accommodate the possibilities. How did a creative person organize his fervor? How did an artist make a life and a living while making art? These questions, but more fundamentally, his inventive drive, cleaved him immediately to his new employer.

"Sam appreciated my aesthetic from the very first day I went to work for him. We had that kind of relationship as well: being fellow artists." Yet Sam saw more than an eager, promising young artist in the guise of a useful helper. "He also saw that I was adrift. His natural character was to reach out."

As a young man, Sam had been mentored by Millard Sheets, the head of Scripps College's art department, who presented a model for how a disciplined artist might pursue a lifetime of creative endeavor. Millard drew Sam into worlds that he could never have penetrated alone—perhaps not even perceived. In a complementary spirit, Sam attended to Larry's yearning to make objects of idiosyncratic grace and particularity. He grasped the young man's instincts, shared his obsession. Sam opened for Larry a portal to the Southern California art world much as Millard had done two decades before. He purchased Larry's early ceramics, seeding a collection that would expand significantly over the years. As a woodworker, Sam demonstrated how to be an artist.

The first six months of Larry's apprenticeship proved arduous. He struggled to understand what Sam desired in the workshop, forced to rely on observation, trial and error, intuition. Famously talky on all subjects except the explicit direction of his work, Sam would only say, *That's a little wide there, that's a little heavy. Take that down. Put your attention there....*

"For the first four years, all I did was sand. When I finished my first set of eight chairs, which took me a month and a half, I put the oil on the first one and it looked dull compared to the others I'd seen in the house. I called Sam over. 'Well, what do you think?' He just stood there and looked at it. I said, 'It needs a lot more sanding doesn't it?' 'Yeah,' he said, 'I think it does.' And that's how I moved to the next level. Sam had very few employees over the years, perhaps a dozen. For the most part, he was hiring creative labor. People with an aesthetic who could understand form. On many occasions, the people involved in the process ultimately affected the aesthetic. Sam was very open. 'If anybody has any ideas,' he'd say, 'let me know....'"

From the start, Larry sensed the opportunity to grow as an artist even as he was learning the basics of his craft. He had already set aside one career. For much of his young life, he'd been a working musician, doubling on tenor sax and vibes, gigging through his teens. His last steady job: a piano bar at the Whispering Winds golf course, a corner of the Tahitian restaurant filled with weekend drunks, the musicians themselves underage and too frequently ignored. By the summer's end, despite his love of music, he knew he didn't want the life. Neither did he seem to possess the facility to write his own tunes. It was 1960, the compositional heyday of Monk, Miles, Brubeck, Bill Evans, Sonny Rollins. Larry stopped playing. To be a musician, music had to be the only thing, and Larry found the world too endlessly interesting. With music, he would simply be replicating the tonal shapes and conceptions of more fluent composers. Imitating, repeating. With art, he could embark on a new language, working it over, teasing out its furthest implications, making it his own.

Five years into his apprenticeship, Larry could finally "take a piece, finish it, and it became my piece. I looked at it, and it was right." By 1969, he had moved on to shaping furniture and running the shop with a small crew of younger men handling most of the sanding and finishing. He also began to grow restless. Larry rented a house adjacent to the shop on property owned by Sam. "My commute was eighty feet. I had a clay studio and a welding studio there. By then, I'd married and my kids were growing up on the property. We were part of the Maloof family—Christmas, holidays, birthdays, all that. But I was very isolated." When his marriage dissolved, Larry determined that it was time to leave. "I wanted to educate myself. I'd dropped out of school to work with Sam. I wanted to get socialized."

A friend at Cal State Fullerton tipped Larry to an opening for a technical assistant in the art department. "It was exactly the opportunity I wanted. I wanted to grow." True to habit and inclination, he threw himself into the job, directly assisting students and faculty with their sculpture, etching, metalsmithing, jewelry, design. Seven years working alongside Sam had readied him to absorb information and technique by imitation, experimentation, osmosis. "All those classes weekly, daily. Printmaking, painting. At the beginning of the semester, I'd sit in on the faculty demos. That's how I taught myself. I learned from observing. Then I'd try it myself when I had time."

He spent seven years at the Fullerton art department and then struck out on his own, opening a small business that drew on his woodworking skills to design and construct artful retail displays. As always, his enthusiasms expanded, mutated, bled into one another. "Any interest I had, I started incorporating them into my work. Landscaping. Building ponds. This was pre-computer, so you had to ferret out information. I paid attention to people who knew things. I was getting wider and wider. A Jack of all trades. I understood that I wasn't a master. That didn't bother me. I was interested in image, and presentation, and perception, and impact. That's what art's all about."

Though it's not what necessarily moves art in the marketplace.

"Art dealers like people who have consistency, continuity. I'm an experimenter. Following the energy that causes images to pop in my head. At first, I couldn't understand how Sam could make the

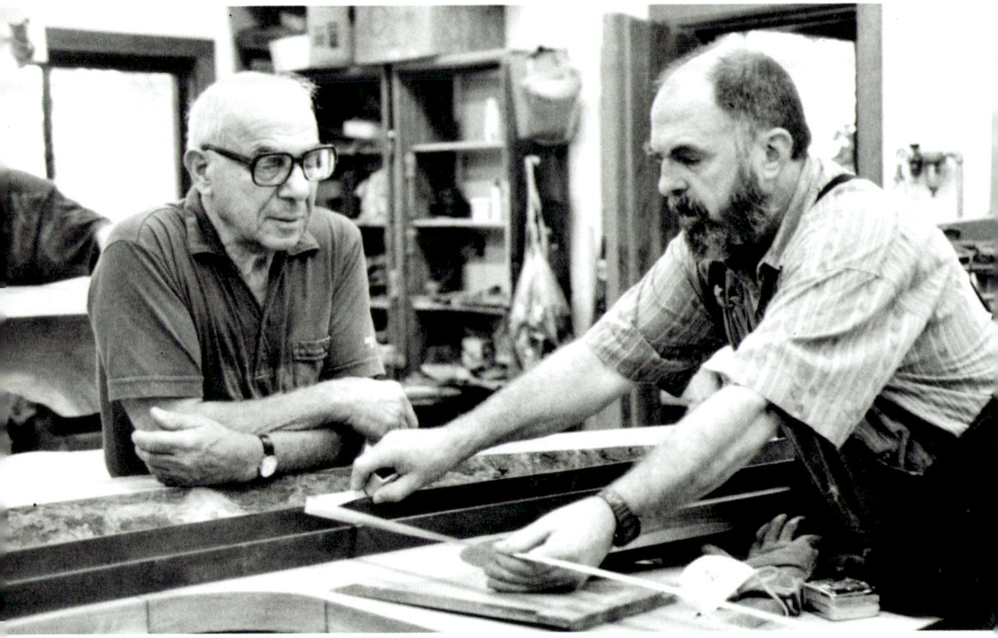

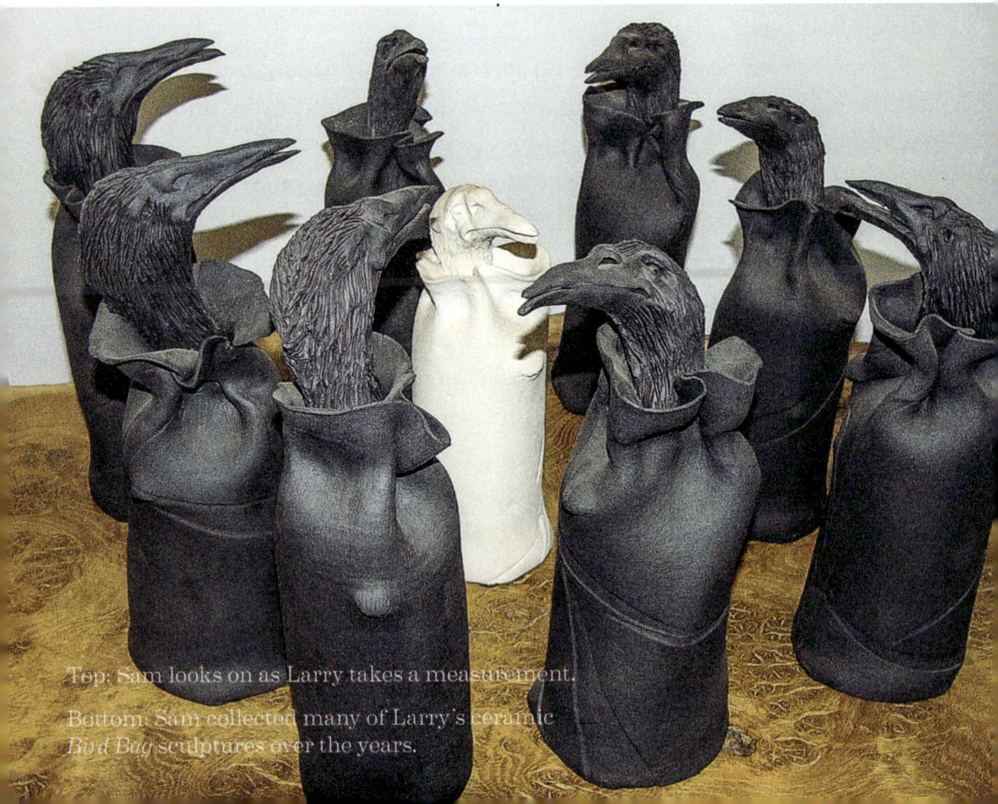

Top: Sam looks on as Larry takes a measurement.

Bottom: Sam collected many of Larry's ceramic *Bird Bag* sculptures over the years.

same thing over and over and over again and be happy. That's not my nature. But Sam loved woodworking. I think about Gandhi and the spinning wheel. Work was his litany. For Sam, coming into the shop and working was how he found a place in the universe. That was his spot."

Sam's ability to concentrate, appraise, refine, and then powerfully refocus his energies to step up to a higher level of design and execution led to what Larry now characterizes as his early, underappreciated achievements—his cabinetry and furniture sets commissioned during the 1960s for select homes and offices.

"Sam was brilliant in case work. I thought that was the real Sam Maloof. It expresses his creative energies. He knew how to design pieces that affected people when they walked into the room. He took the breath out of them." But over time, fewer clients elected to hire a single designer for an entire suite of furniture. "In the sixties," remembered Larry, "we'd do a whole room, sometimes both the dining and living room. When the customer ordered a table it included eight or ten chairs. A set." By the 1980s, wealthy buyers were collecting contemporary furniture's big names rather than allowing a lone designer to articulate a complete and cohesive vision. "You'd buy a Sam Maloof table, and the chairs from somebody else. That aesthetic affected the market. I noticed a trend."

In March 1992, after a quarter-century absence and a decade-long discussion regarding the prospects for his return, Larry went back to work at Sam's shop.

"When I left, Sam was not happy," acknowledged Larry, "not pleased. He thought I abandoned him. We didn't see each other for a couple of years." Then, during the early 1970s, Cal State Fullerton's art department produced its *Chairs in Motion* show. Sam agreed to be a jurist. Larry entered a potentially contentious piece—a rocking chair. "Sam saw that what I had done with the rocker wasn't anything like his, so he didn't feel like I'd run off with all the juice. We walked around the show together, touched base again. The next year, I taught furniture design and advanced crafts. I called Sam and asked if I could bring my students to the shop. That broke the ice. We reestablished our friendship. Then, over the years, I'd visit, stay in the guest house."

The two men's levels of experience and status had struck a more equable balance, their interests and approaches intersecting, veering off, pooling and comingling. Larry settled back into his "day job"—a half-jesting acknowledgment that woodworking once again provided the steady paycheck and emotional ballast needed to extend himself more riskily as an artist. And once again, surprisingly this time, Larry found himself struggling to keep up with Sam's designs.

"The furniture was completely different from what I'd worked on in the past. Much more complicated sculpturally. I had to learn how to shape the new pieces." Mike Johnson, who had been working alongside Sam for eleven years, assumed the tutor's role. "He was a good instructor. But Mike and I don't work alike. I work like Sam did. I'm more intuitive, more sensing. I don't measure as much. After a period, I abandoned a lot of the techniques and started working the way I always did."

Larry's return to the shop after twenty-five years also provided a telling perspective on Sam's early pieces. "They were very complicated forms to make," said Larry. "But easy to sand. Over the years, they became much more logical in their construction. Sam could literally put a low-back together in two days. But it now took much more time to shape, hold the lines, make it work aesthetically, and then finish and sand completely."

Repairs, too, prompted an inventory of past concepts and techniques.

"When I first started working in 1960, Sam wasn't sanding the bottoms of chairs. I didn't know any better and it bothered me, so I started sanding out the bottoms. He liked that. When we get a really early piece in these days, we flip it over. 'Yow! Check this out.'

It's really cool to observe how Sam grew. He started out knowing nothing, self-taught. Then, eventually, he hired people who affected the look, helped make it grow."

Today, the evidence of Larry's contribution to the Maloof enterprise can be viewed throughout the Maloof property. There's the garden space allotted to his iron-and-timber balance beam. The fan of wooden shafts recycled from Sam's original work site and refashioned into yard sculpture. The three dozen ceramics, old and new, stationed throughout the residence.

In 2014, Larry officially retired, relocating from his longtime home in Desert Hot Springs to a house in the foothills past Pioneertown, nearly two hours' drive from Alta Loma.

"Something I've been dreaming about for over twenty years showed up suddenly in the high desert. It's got a big thirty-by-thirty studio. When I walk through that door, I feel this glow inside me that I haven't felt in a long time. I'm going to be absolutely rejuvenated. I already am. I just turned seventy-two years old. I'm looking at the clock. When I left Sam's shop, I thought I'd go for a career, get representation, try to get into galleries. Then I suddenly had this realization—that's not at all what I'm going to do. It's not what I've ever done. I'll be making art. The ambition part has nothing to do with me. I just want to make things. I'm plagued with it."

2.

MAKING FURNITURE
MIKE JOHNSON

January, 1981. Mike and Joanne Johnson were shopping upstairs at Penny's in Montclair, not far from their tiny newlyweds' cottage in San Dimas. Mike pointed across the floor to a couple in their midsixties. The man was someone he had often discussed with Joanne: the woodworker Sam Maloof, a master craftsman—an artist, really. He lived just a few miles away. Imagine....Mike had idolized him for years.

Then Mike stepped onto the escalator, intending to stroll out into the parking lot and make his way back home. Joanne reached for Mike's arm, pulling him back.

"You should introduce yourself," she said. "You'll regret it if you don't."

Mike still remembers the expression on the face of his nineteen-year-old bride. An indication of utter seriousness. Even then, thirty-four years ago, he understood precisely what that look meant.

As they turned around, they could see that Sam and Alfreda had been watching them. Sam stepped forward, graciously introducing himself. Mike told Sam that he had followed his work for years, most recently attending a lecture Sam gave in Los Angeles during his final semester in school. Mike had just finished his bachelor's degree in industrial arts at Long Beach State University. For now, he worked odd jobs for his landlord—painting, cleaning up. But he aspired to become a woodworker himself, setting money aside to purchase basic equipment so that he might start his own business.

Sam invited Mike to the shop. Mike was elated, Joanne vindicated. The very next day, he packed up his truck with several pieces of his woodwork, hoping to demonstrate his skill and commitment to the craft. He already knew what to expect when he arrived at the Maloof compound. One of his instructors had screened a series of slides about the property and handcrafted house.

"But when you get here," said Mike, "and you're toured by Sam, you have a different kind of insight. This is the man's life, his environment. The whole lifestyle, the way that he works and thinks about how he works. He's not aiming to make a million dollars. He's trying to fulfill something in himself." The house, grounds, and workshop, the artwork hanging on the walls and nestled into every corner, Sam's blend of geniality and earnestness, his percolating energies, the primacy of beauty and utility at every turn—"it was all a package that goes together."

It also became evident that Sam harbored no curiosity whatsoever about Mike's woodworking ability. He nodded with polite indifference at the pieces stuffed under the pickup's shell. Instead, they chatted.

I think he was interested in who I was as a person and if we were compatible. We talked for quite a while. Finally, he said, 'I'd like to have you work for me.' That totally took me by surprise. I started the next Monday. It was February 2, 1981."

Today, Mike Johnson runs the Maloof shop with the help of his son, Stephen. In the course of three decades, he has become a skilled and respected craftsman, his career shaped by and fused with the progress of Sam Maloof's designs, techniques, practices, and philosophy.

"I'm still interested in what Sam was doing and how he was doing it differently than most everybody else," said Mike. "For one thing, most people when they do joinery, they're interested in doing it so that the joint is hidden. Sam's philosophy was when you design a good joint, it should be out there for you to see, appreciate. Why go to all the trouble of making this intricate joint that's beautiful and highly functional and then hide it?"

How this might be achieved was communicated by Sam in subtle fashion.

"There wasn't a lot of verbal instruction," acknowledged Mike. "He'd say, 'Here.' And he'd grab your rasp. 'Let's do it this way....' Oftentimes, there were just a few words, and then he'd let you run with it and come back to critique."

What Sam communicated most powerfully by way of his attitude and demeanor, through physical gesture and a running conversation about everything other than woodworking, was a set of values that might be deemed conservative. Discipline linked with clear intention. A premium on balance, stability, precision. Sustaining traditions, even traditions of one's own making, while pursuing improvements that proceed gradually toward a possibly unrealizable ideal.

Sam taught his young apprentice "how to master exactness in this craft when most people think close is close enough. Perfectionism," said Mike, identifying a quality that feels untroublingly blended into the best Maloof creations. "Some people might say that that kind of philosophy makes it hard to live with a person. It's almost a sickness. But you have to know when it's appropriate. Sam had control of that. He was a model for me."

In matters unrelated to woodworking, Sam also exerted influence. "He preached to us that when it came to money, don't spend anything you don't have. He wasn't a believer in credit. Sam would use credit cards as a convenience, but he always paid them off. That has become an example for me and my wife."

Today, with Sam's early pieces of furniture commanding staggering prices and his professional regard inclining toward veneration, there's a temptation to draw larger inferences from the work. Perhaps the visible joinery of a Maloof chair contains a message about how we are allied in human society—the individual pieces of our separate lives fit together in families, communities, culture, and civilization. After all, don't we expect art to instruct as well as delight?

"If that question were posed to Sam," replied Mike, "I think he'd say that you were reading in more than was necessary. There were always people who were trying to dig deeper into what Sam was doing. Or trying to figure out the formulas for his design, his proportions. He would say, 'I just make it because it looks good. Or because it fits good.' He would never, ever admit that there was anything more to it. I've never been highly philosophical in my thinking, either. But I'm not going to say there's not a subconscious level that influences us."

Then might not the creative drive be founded on the need to make palpable what we don't understand or can't put into words? Isn't it a fine thing to introduce into the world an object of beauty and use, and let it speak for itself?

"Sam's work has always been pretty simple," insisted Mike, "straightforward. You get what you see. Yet what you get and what you see is enough. Maybe you don't need anything more. If you can be satisfied on both a visual and comfort level, maybe you don't need to look any deeper."

Recently, the work generated by the Maloof shop has involved a good deal of looking back. At least a quarter of the jobs, calculated in terms of time and revenue, involve restorations, repairs, and refinishing of Sam's earlier work.

"This year alone," said Mike, "we're approaching thirty separate restorations, some with multiple pieces."

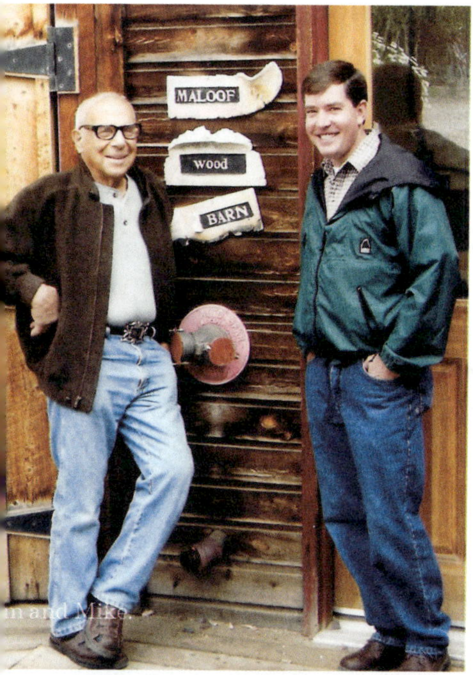

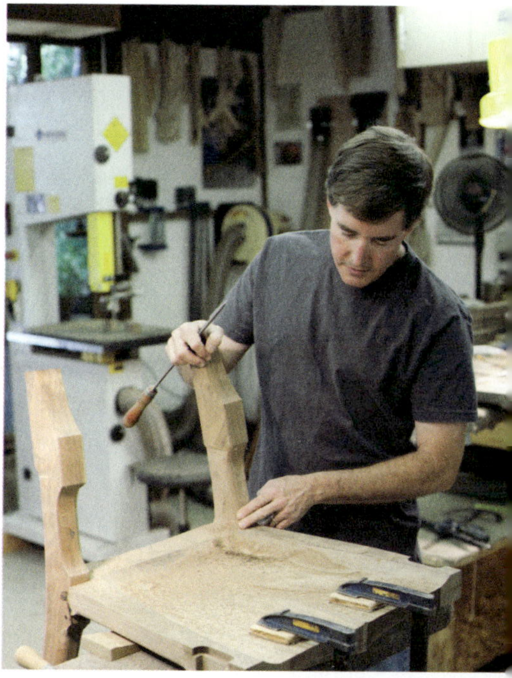

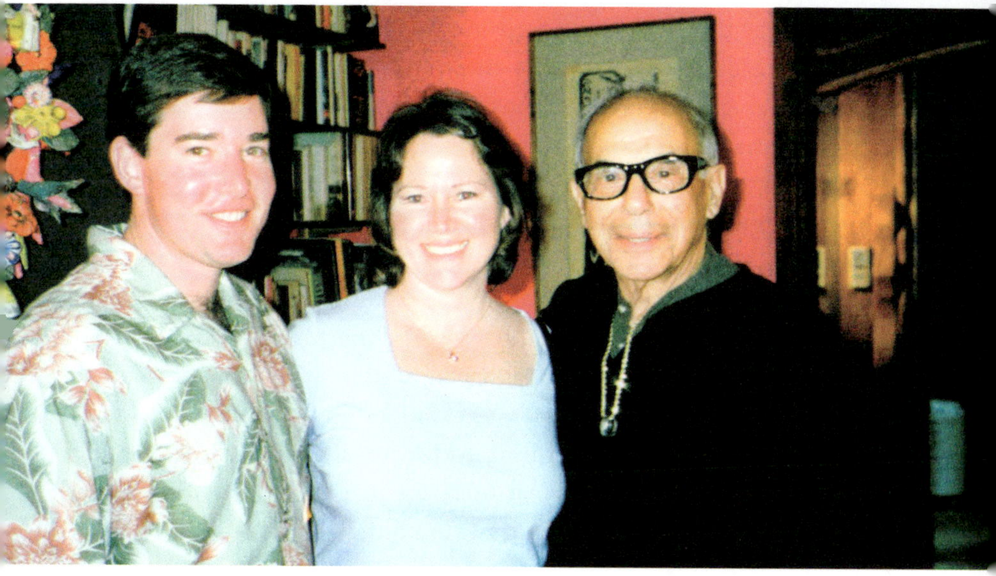

By 1990 when this photo was taken, Mike and his wife, Joanne (center) had been part of the Maloof circle for a decade.

Chairs and tables arrive for mending before hitting the block at Los Angeles auction houses, the surfaces sanded and refinished to quicken the pages of their glossy catalogs. Other pieces are passed from one generation to the next, the new era of ownership commemorated with repairs and reupholstering.

"We're going to pick up a raised conductor's platform with a rail that we built in 1996. It's used by the Redlands Symphony, and they now want us to lower it. We'll cut it down and refinish the whole thing before their fall concert series. We just repaired a communion table from a church in Altadena. When it left here, it looked like it did in 1967 when it was built." Occasionally, Mike and Stephen make a house or office call and re-oil pieces on site, though he prefers to labor in his own noise-and-dust-tolerant shop far from observers who lack the patience to watch finish dry and allow barely perceptible layers to accrete over many coats.

"It's humbling for me to work with Sam's old furniture," said Mike. "To see pieces that were built before I was born. By the time I started working for him, Sam was sixty-five years old. Most people would be thinking about retirement. He was beginning the second half of his career, from sixty-five to ninety-three."

A thought experiment that Mike finds instructive: Place side by side Sam's first and last pieces. The differences glare back at you; they're extreme, jarring. Now find yourself an enormous building—say, one of the cavernous display halls contained at the LA County Fairgrounds—and line up all five thousand–plus pieces of furniture and related woodworking, the sum of Sam's career. As you amble from piece to piece, inspecting the planes and curves, the joints and latches, the strategies for assembly and the regimens for finishing, you would witness an understated evolution, as though progressing steadily, if minutely, toward a Platonic ideal—the archetypal rocker, the perfect table for use in its time. Sam's work never stopped changing, and never changed abruptly.

"Just yesterday, I delivered four chairs to Claremont McKenna College," said Mike. "They had me replicate a conference chair that Sam made in 1967"—one of twenty-five fit around a boardroom table that Sam also built. "The trustees had sat countless times on these chairs, so they thought what better piece to replicate? But in 1967, I was nine years old. When I started with Sam, the work had evolved to a point where everything was considerably different in construction, joinery. It was much more sculptural."

Even though the assembly method proved relatively simple, Mike "had to get into Sam's head" to grasp its strategy.

"Being with Sam as long as I've been, working with him on a daily basis, you start to understand how he thinks. I'll ask myself, 'How would Sam approach this job?' I picture what I've seen him do over and over again. Then when we get pieces back that are broken, that's the best education. I can see it from the inside out."

At the moment, Mike and his son, Stephen, are finishing repairs on a coffee table built sometime around 1964 and damaged by the owner. "She fell on it, and broke the table's leg. Luckily, she didn't

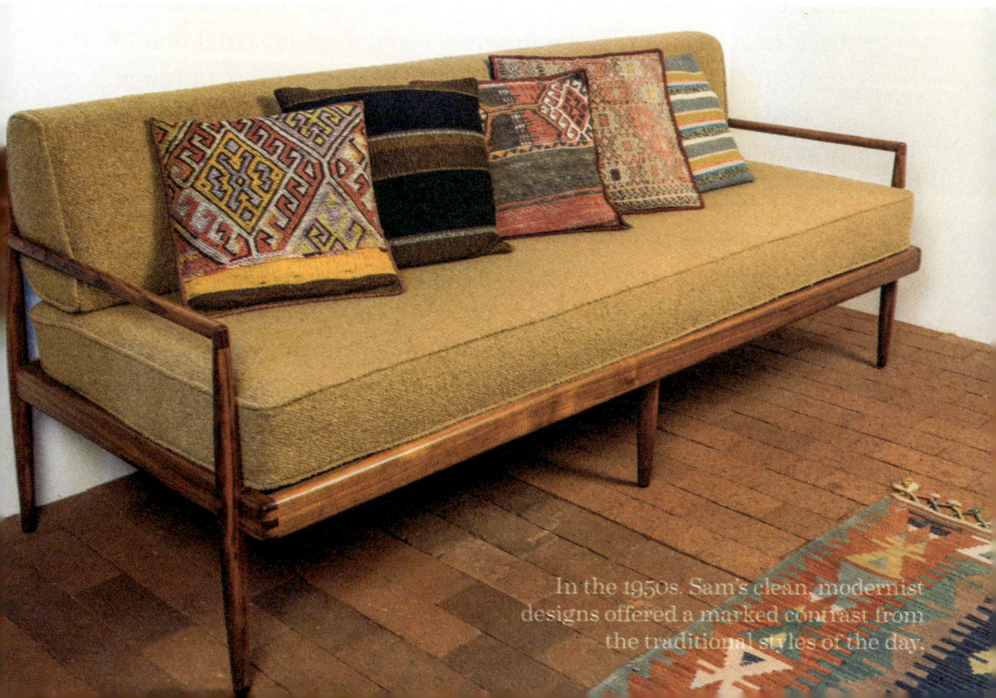

In the 1950s, Sam's clean, modernist designs offered a marked contrast from the traditional styles of the day.

get hurt at all. We picked up the table, repaired it, even made it stronger. I've learned how to reinforce early joints by strengthening the legs, placing the screws in such a way that they're hidden. Techniques that Sam maybe didn't feel were necessary, but time has proven them to be." Today, many repairs leave the shop stronger than when they arrived.

"It's good for me to see these older pieces," affirmed Mike. "To see not only how they were built, but how people use them. When a piece left the shop, we'd often forget about it—we'd be on to the next one. I often feel like we're in a cave here where we don't see what the outside world is doing with Sam's furniture."

Still, Mike's devotion to a single shop over four decades has delivered him to many places otherwise inaccessible. A billionaire's mansion. A workshop in Maine where craftsmen hurriedly assembled the staterooms for a 149-foot yacht, later to be furnished with twenty-five Maloof chairs and tables. Just last year, Mike spent seventeen days in a remote Kenyan village, organizing a vocational woodworking program for an orphanage sponsored by his church. His son joined him, making the six-hour trek to the nearest facsimile of a hardware store on three arduous occasions and bringing back tales of animals slaughtered, cooked, and feasted upon in the same alleyway—visions incalculably distant from Alta Loma.

Mike also traveled to Washington, DC, for Sam's retrospective at the Smithsonian's Renwick Gallery—the *King of the Rockers* show featuring seventy-six pieces of furniture, slated to open the second week of September 2001.

"We were there on 9/11. The guards at the Renwick were standing in front of the television set as we tried to figure out what they were so intently watching. It wasn't long before we could see smoke from the Pentagon. We heard the air force jets take off to pursue the plane still in the air above Pennsylvania. We couldn't see them, but they made a loud sonic boom." Naturally, the Smithsonian postponed the opening. With flights grounded and the Capitol nearly deserted, Mike and Sam toured the museum in eerie, delightful emptiness, slipping in and out between sporadic high-alert evacuations. "We have photographs of us, the only ones there when it's usually packed. What I'm trying to say is, I've had experiences because of Sam that I would have otherwise never have had...."

Experiences that had to be earned gradually, patiently.

"After years, there came a time—and I can't put a date on that time, you just knew it had happened—when things changed. I knew I'd crossed over." No longer a student or helper or apprentice woodworker, Mike quietly assumed a position attained by few. He had earned Sam's confidence.

"I've always been very grateful for this job," Mike emphasized. "I'm honored to be a part of it. We're making heirlooms. People will pass these pieces on. And while my coworkers over the years—especially Larry—were first of all artists in their own right, I never thought of myself in that way. I thought of myself as Sam's craftsman. And very content at that role."

Craftsman. A term of modesty and immense achievement.

"You can argue all you want," said Mike, "but if it wasn't for the furniture, none of this would be here. Sam started in a one-car garage in Ontario making furniture. That's what his reputation is built on." Mike gently splays his fingers—a diffident, barely indicative gesture that has to encompass the house and gardens, the museum, the art collection, the visitors now arriving to tour the grounds. "All this has become part of it. But the fact is, he was a furniture maker."

3.

TABLE
HIRAM CHODOSH

"Every piece of furniture puts you in a slightly different posture," explained Hiram Chodosh, president of Claremont McKenna College, about nine miles from the Maloof shop. "In the grander sense."

That favorite chair from childhood subsumes you in a warm bath of Lay-Z-Boy reverie. Your desk at work commands attention (while the afternoon creak in your sacroiliac reminds you to find another job). The dining room table, inherited from your mother who prepared the pot roast in a manner you've never been able to duplicate and overpopulated each flank of mahogany with a boisterous delegation of warring uncles and aunts—that particular piece of furniture evokes the full jumble of feelings associated with cantankerous spirits of the past.

We don't usually connect furniture and emotional complexity. But consider the root, from the Old French: *fournir*. To supply, to give. To equip with what is needed. How we respond to the necessities furnished for work and home determines to an underappreciated degree how we feel about both.

"If you walk into a classroom," said President Chodosh, "and it's got a big lectern at the bottom of an amphitheater and all the seats are directed at that person, what does it tell you about the learning process? If you walk into a boardroom and you're scared to put down a cup of water for fear that you might destroy something, and the seats are rigid and uncomfortable, and the people are situated at too much of a distance from one another—then just what does it say about the ethos of that particular organization? For me, a piece of furniture is the shape of a social process."

Which led President Chodosh to admire the virtues of Claremont McKenna's conference table. On the second floor of the Bau-

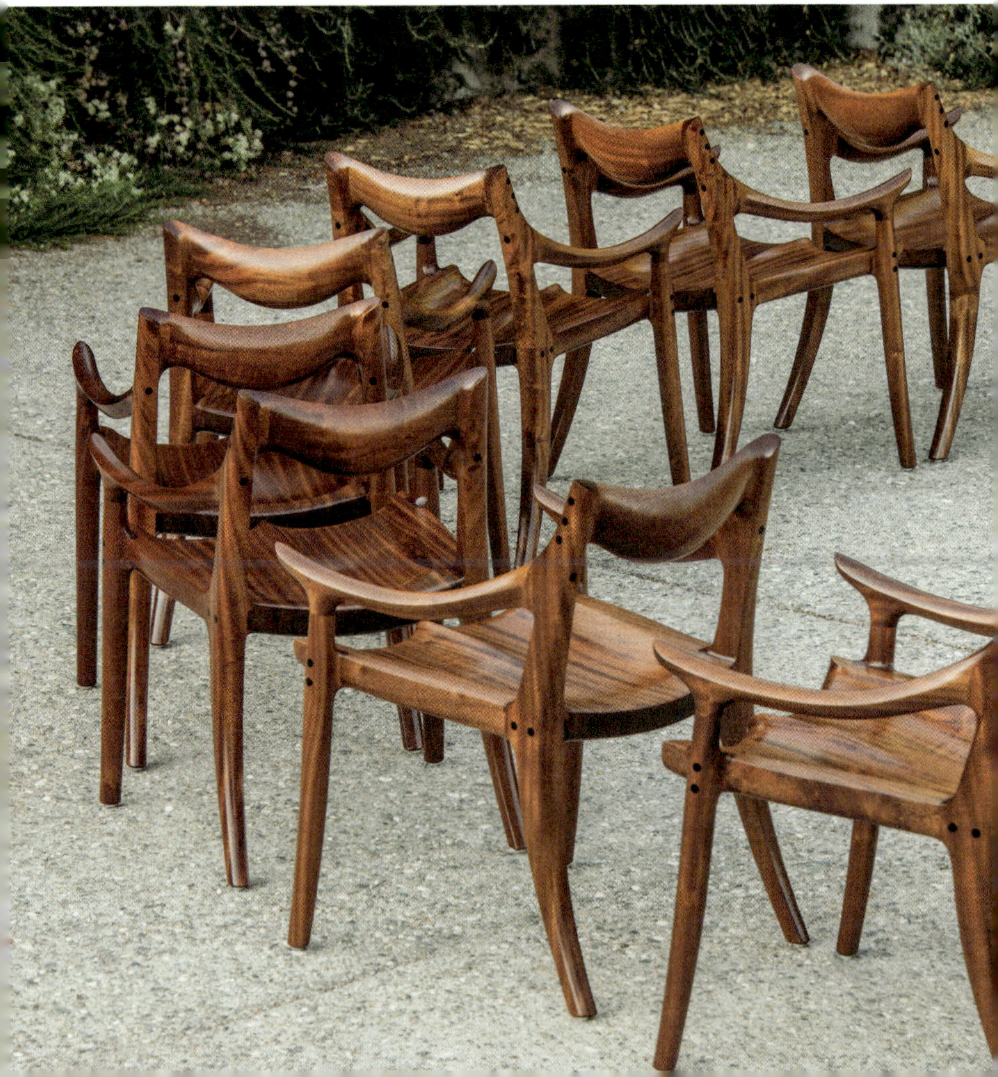

er Center administration building, a few feet from his own office, resides an eighteen-foot-long table dedicated to the equable deliberation and decision making of the university's trustees. It's a gorgeous piece, an early work completed by Sam Maloof in 1967. Walnut base, African Shedua surface, a set of black walnut chairs encircling the perimeter.

"It's solid," said Chodosh. "It's wholesome. It has a certain gravitas associated with the choice of wood and the design. On the other hand, there's a subtle elegance that doesn't call attention to itself. That's one of the keys to a great boardroom. It carries enough weight so that people get a modest sense of awe when they walk in. But it's also sufficiently comfortable and intimate so that new ideas and conversations are possible."

Some days, four or five administrators will huddle at one corner to sort through subjects whose significance or sensitivity requires a formal setting. Other times, the full board of trustees assemble to discuss policy, ultimately making decisions that will direct the course of education for the small college's 1,300 students.

But does the table really matter?

"What's great about Sam's furniture," explained Chodosh, "is the knowledge that each piece is handcrafted. And that creates a conversation between the user and the furniture that can't take place when the furniture has been slapped together in a factory and sold as a commodity, instead of an experience." The character of the piece, its negotiation between functionality and beauty, provides "a feeling that great ideas and decisions have passed across the table in that boardroom. It encourages people to think freshly about old problems and new challenges."

The table raises the stakes.

"With Sam's craftsmanship," concluded President Chodosh, "we all feel that we have to rise to a similar level."

4.

OXYGEN
KATHRYN GOLDEN

Kathryn Golden's family moved to Alta Loma in time for her to enter the sixth grade. She remembers a speck of a town with the population totaling perhaps 2,500: A single key intersection, a volunteer fire department, a closed café, a bar still doing a fine business. There was one gas station, a small library, a tiny market called Billings', a beauty parlor, and the school running from kindergarten to eighth grade.

In the sixth grade, Kathryn met Slimen Maloof, whom she would later marry at the age of seventeen, immediately after high school. They are no longer married today, but would spend the next twenty-five years living on the Maloof family compound—raising their children, mixing in an expansive swirl of friends and fellow artists, working in wood and leather as they pursued their own lives, while Sam's reputation grew from small-town artisan to world-famous woodworker.

So, of course, Kathryn remembers the first time she saw Alfreda—then imaginable only as Mrs. Maloof.

"She was walking into the Billings' market," recalled Kathryn. Alfreda wore a long, full orange skirt and a Southwestern silver and turquoise concha belt, her hair heaped atop her head. "This really striking woman." Her "air of grace, uniqueness, and almost otherworldly presence" seemed magnified by the distance—Kathryn stood about two hundred feet away—as well as the pronounced height of the sidewalk, a two-foot-high curb that served as a wash for Cucamonga Peak's runoff about two miles to the north. Stepping from the sidewalk into the store, Alfreda appeared elevated, as if gliding across a stage.

As a child, Kathryn didn't have the words or experience to clarify what she observed or felt, but recognized that Alfreda exerted a powerful draw. Beautiful, quiet, subtly mysterious—a portrait of the artist as a young mother. Room Mother, in fact, for Kathryn's elementary school class; Alfreda visited one day to set up a model pueblo and talk about her years spent teaching in the Indian Service in New Mexico and Montana. She had already had a full, adventurous life before marrying Sam. Fresh out of UCLA with a teaching degree, Alfreda wanted to become an artist, but the Depression intervened. She took a job with the Bureau of Indian Affairs, assuming responsibility for a classroom at the Santo Domingo Pueblo, eventually transferring to Santa Fe as a teacher of fine and applied arts at the Indian School. She made repeated visits to the Pima and Papago reservations to investigate Indian crafts, and she spent two weeks at San Ildefonso Pueblo alongside the famed potter Maria Martinez, who became her lifelong friend, and years later, Sam's friend too.

Around town, Kathryn had noticed the Corvair truck driven by Sam for furniture deliveries. It bore a large M decal on the side with the Maloof insignia. "That's a business truck," she thought, "but it isn't a business truck." Just right for a business that proved to be far more than a business. Right for Alfreda, too, with her gentle, sturdy way of not quite fitting in with the common culture of greater suburbia that had begun to encroach upon the borders of tiny Alta Loma.

"The best thing about her," said Kathryn, "was her spirit. She exuded peace and intelligence." Kathryn and her children called Alfreda "Boppi" and Sam "Boppa," the nicknames assigned by her son Aaron as a toddler. "I loved them both," she said. "And it was mutual." To friends and admirers, Alfreda appeared the model of balance and poise. "Sam would be bustling around," remembered Kathryn, "because Sam was always bustling." Alfreda was calm and organized, her steadiness exacting a gravitational pull on the family and the business. Her humor was dry and understated, her elegance enhanced by a slight reserve.

In the estimation of friends who intimately knew them both, Alfreda proved the indispensible partner. Through her exertions and patience, she had enabled her husband to become Sam Maloof, renowned woodworker, world-famous artist. Sam acknowledged as much in the opening pages of his autobiography, thanking "my wife, Alfreda Ward Maloof—without her love and help nothing would have been accomplished."

What did she do?

At the start, Alfreda urged Sam to pursue his ambition of making a living at making furniture. "If you want to try it, do it," she said. "We can manage." She ran the business, watched the books, handled the sales and correspondence, mined and maintained the mountainous records. Cared for the children, supervised the calendar, fed and entertained the guests. "I remember her in the kitchen," recalled a frequent visitor. "She was always preparing food. As Sam became better known, people made pilgrimages to the shop. He'd talk to them and she would feed them." Lunch was nothing fancy—cold cuts, sandwiches, yogurt churned into a batch of cottage cheese—but each dish was composed and presented with a grace note of consideration and care.

"She made everything from scratch," remembered Kathryn. Salads every night with the dressing brightened by lemons

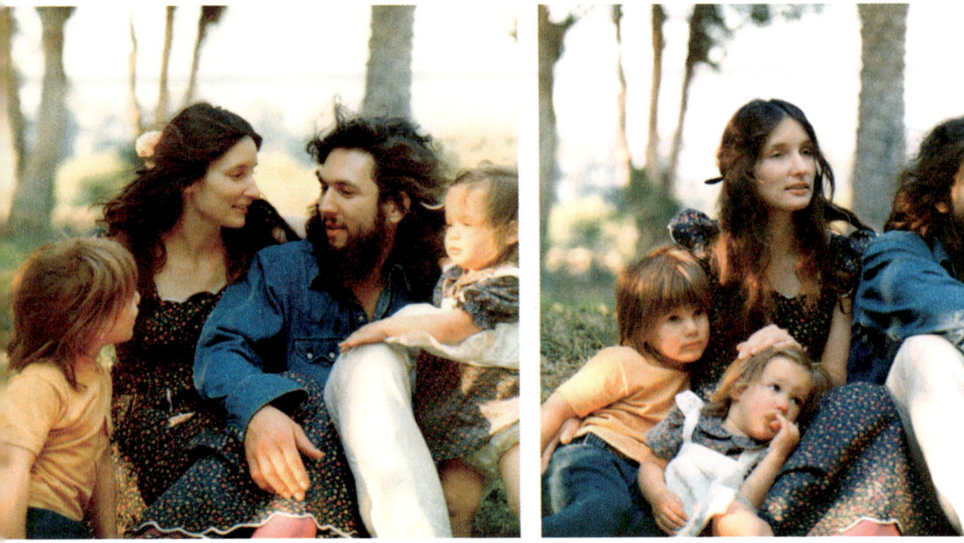

Kathryn, Slimen, and their children, Aaron and Amy Rebecca.

picked fresh from the grove. She added olive oil and ground the herbs with a mortar and pestle. "She used cilantro before many people had even heard of it. We had a Sunday afternoon family meal that was generally a roast of some sort." They assembled around Sam's cork-topped table, each member in their designated place. Sam would always ask for bread. "Do we have any bread, Freda?" Sam always called her Freda, not Alfreda which she preferred. Alfreda was famous for her delicious pickled figs, an accompaniment to meat dishes, the figs plucked from a large tree on the grounds. She cooked with nuts, dates, whatever fruit had ripened. Made fresh lemonade when the temperature soared. "She made a delicious tuna salad on mixed greens, avocados from the trees, persimmons or oranges, red onions, and sourdough baguette sliced and spread with butter mixed into mustard and sprinkled with za'atar and toasted in the broiler. Perfect!" Kathryn remembered. "She made Swedish rice pudding as comfort food for family and friends who needed some comfort.

She could cook anything. She was a whiz with vegetables and spontaneous sumptuousness."

Then she nudged Sam back to work, preempting visitors from tarrying too long. Before Sam employed an apprentice—sixteen years into his career—Alfreda often lent a hand in the shop, his sole assistant. She conferred and strategized with him over plans for the future, supported his aspirations and kept him on track. Alfreda sustained and nourished the family, the business, the possibilities of making a living and a life from art and craft.

"'If it wasn't for Freda...'" one old friend remembers her mother remarking, just a bit ruefully.

"It would be any artist's dream to have somebody like Alfreda," agreed Kathryn. "Somebody to take care of everything so you can pursue your work. She was just a perfect person."

Then why does any characterization of Alfreda's role or enumeration of her contributions finally seem inadequate?

In the complicated progress of a fifty-year partnership—an immensely happy, satisfying marriage by all declarations and accounts—some imbalance of visibility and regard seems almost foreordained. Sam wielded his band saw with grace acquired over decades; Alfreda made those years feasible. And today the world admires the career and creations of Sam Maloof.

Contributor, collaborator, partner, wife: all seem weak and inestimable reductions of the truth. More accurate, perhaps—more just—to consider the moment of vivid absence, Alfreda's sudden death in 1998. The loss undermining the basis of Sam's world, his closest friends fearing that he would soon follow. Alfreda was the atmosphere that supported Sam's career and breathed possibility through both of their lives, the oxygen one might not even notice until it vanished.

5.

PEACEABLE KINGDOM
HARRISON MCINTOSH

Harrison McIntosh and the interviewer from KCET, the local PBS affiliate, position themselves on Maloof chairs at the Maloof table. Behind them hangs a large abstract painting by Karl Benjamin. Just out of the camera's eye you can spot another abstraction by James Hueter, his oil-on-wood *Grey, Flush, and Setback.*

While the four artists spent many, many hours in one another's company (along with Rupert "Rummy" Deese, with whom Harrison shared a studio for fifty years, whose ceramics are huddled slightly off camera), this particular arrangement of their artwork serves as a kind of mockup, a symbolic recreation of mid-twentieth-century creative life in the Pomona Valley. The occasion is Harrison's one hundredth birthday party, marked by a grand commemorative exhibition at the American Museum of Ceramic Art in Pomona, California.

Four hundred friends and associates mill through the galleries, pausing at Harrison's early terracotta torso, the glazed stoneware illustrated with Mishima circles and lines, the vases and plates, lidded jars and footed bowls, the bottles and chrome-base sculptures. A balance of pattern and movement characterizes each display. "I always had classical music playing on the radio when I was working," said Harrison. "I feel my work was close in kin to music, especially Bach." The exhibition confirms this notion: the repetition of line and figure, the energy and serenity, the precision of detail culminating in a state of tremulous equilibrium.

Harrison's admirers span the globe, his ceramic vessels collected in more than forty museums, from the Los Angeles County Museum of Art to the Smithsonian, the Louvre, and the National Museum of Art in Tokyo. Over the course of a half century, he has been the focus of forty-three solo exhibitions in Los Angeles, San Francisco, Chicago, Baltimore, Detroit, Wichita, and beyond. A generous portion of this achievement Harrison attributes to the happy accident of moving to the Pomona Valley at a moment when a contingent of like-minded artists had caught a similar drift.

By 1945, the war had ended and the possibilities of the Servicemen's Readjustment Act—commonly known as the G.I. Bill—were just beginning to dawn on young veterans returning from Europe and the Pacific. The G.I. Bill covered tuition and a living stipend, making it possible for individuals who had never considered the possibility of higher education to enroll in a university and radically alter the course of their lives.

After being discharged from the army, Harrison entered the newly established Claremont Graduate School as a "special student," a nonmatriculating resident specializing in ceramics. Other fledgling artists joined the ranks, some earning degrees, others spending months and then years in workshops and studios until their benefits ran out. Students included Paul Soldner, recently discharged from the army after helping to liberate Mauthausen concentration camp, later to become a pioneer of American raku pottery. The sculptor John Svenson, an army air corps aircraft mechanic and native of the Pomona Valley who studied with English-born Albert Stewart. Former B-24 turret gunner and painter Roger Kuntz. Painter and sculptor Jim Hueter, a former army camouflage artist, and ceramicist Rummy Deese, an army air corps mechanic. The veterans brought to their studies an intensity and highly compressed ambition, reshaping the campus culture. "An eighteen-year-old out of high school," observed navy vet and painter Karl Benjamin, "is a lot different than a twenty-one-year-old with three years of being shot at." Studios remained open twenty-four hours a day, and the young vets, with vivid memories of the European front and the South Pacific, poured themselves into their work like beneficiaries of a mortal reprieve.

The indispensible man behind all this activity proved to be Millard Sheets. Under his leadership, the Scripps art department had attracted a distinguished faculty, including painters Phil Dike and Henry Lee McFee, designer Jean Ames, ceramicist William Manker, and Albert and Marion Stewart in sculpture and weaving. Millard seemed to be everywhere. He oversaw the commission of a one-hundred-foot campus mural by Alfredo Ramos Martinez, one of the era's leading practitioners from Mexico. He organized first-rate art exhibitions at the Los Angeles County Fair. *6,000 Years of Art in Clay* brought to the region 670 pieces from Egypt, the Middle East, Greece, Rome, and the Etruscan age. His *One World of Art* show controversially displayed work from the ex-belligerent nations of Japan, Germany, and Italy shortly after the war. In 1953, the fair, under Millard's direction, exhibited the advance guard of nonrepresentational art with canvases by Robert Motherwell, Hans Hofmann, Adolph Gottlieb, Ad Reinhardt, and William Baziotes.

Millard also designed and executed a series of mosaic exteriors for local and regional branches of Home Savings Bank, hiring assistants from the pool of Claremont vets. Harrison participated in the project—along with Sam Maloof, who had served in the army but never used the G.I. Bill to enroll in college. Sam benefitted in other ways from Millard's presence. For four years, Millard employed

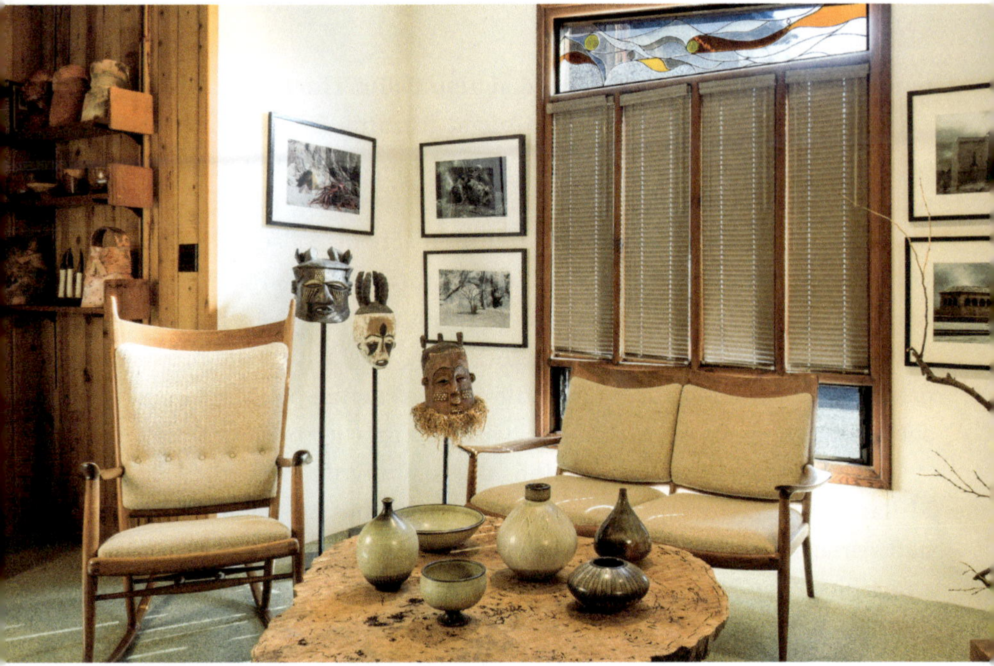

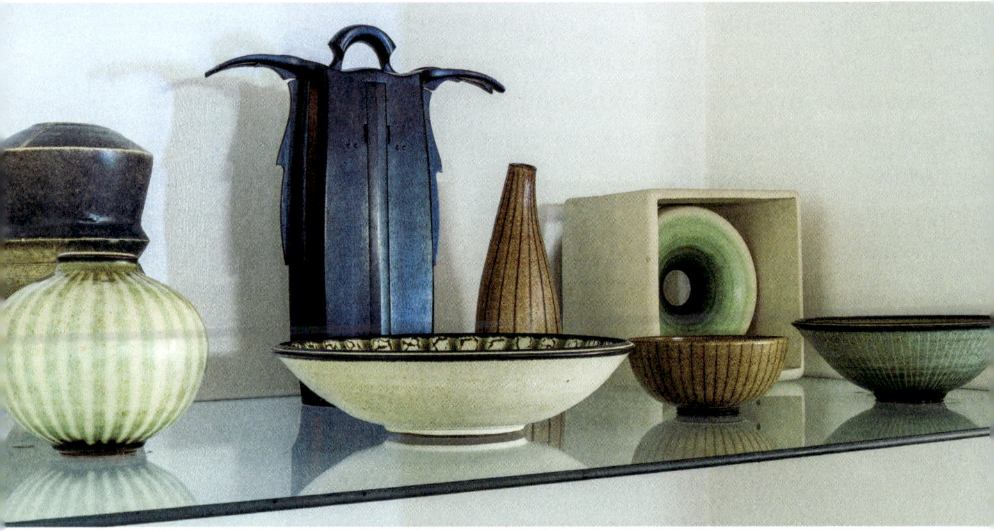

Top: As the Maloofs' art collection grew, Sam added areas for display, including the central mezzanine space. The three vessels in the group on the left are by Harrison McIntosh. Bottom: In the back row, the two ceramic vessels on the left are by Larry White; the rest are by Harrision McIntosh.

him as his assistant, collaborating on a myriad of construction projects at the Sheets' residence, making signs, and building picture frames. Most important, Millard introduced Sam into the university art scene, where the gregarious young man threw himself into the task of making friends.

Claremont was then a small town of three thousand people—a village suffused with bucolic intimacy and buffered by distance and tradition from the surging population of central Los Angeles and its surrounding suburbs. In contrast to the dominant New York art scene, where the blistering urgency of the abstract expressionists' new paintings often matched their frantic, disorderly conduct, the young Pomona Valley painters, sculptors, ceramicists, and woodworkers struck a more conciliatory, balanced tone. "We learned a lot from abstract expressionism," said Harrison—though he and his friends rejected its design for living. Many of the young vets were married, starting families, preparing themselves for careers. On the heels of history's great nightmare, in an act of will and forgetfulness, they immersed themselves in a vigorous creative discipline to achieve a kind of frontier bohemian normalcy. If they adhered to any discernible aesthetic philosophy, it might have aligned with Flaubert's dictum: "Be regular and orderly in your life, like a bourgeois, so that you may be violent and original in your work."

Order brought freedom. Harrison, Sam, Rummy, the entire corps of young artists, enjoyed the autonomy to devise their own interpretations of the artist's life. No school or method prevailed; there wasn't a Claremont style. Under the shield of Millard's entrepreneurial zeal (his character seemed better suited to a field general than a department head), all approaches to art could gain a foothold, as long as they registered exacting and true. Disciplines mixed freely. Ceramicists, painters, woodworkers, and weavers relished the camaraderie, exchanging perspectives, techniques, and materials, and eventually trading among themselves their own creations: a painting for a pot, a chair for either. Friendships ignited and burned on for decades. "We all felt like we were doing something new," said Harrison. "We enjoyed seeing the results of what each of us was attempting. We supported each other, admired one another's work."

Harrison and Sam began eating lunch together at Walter's in downtown Claremont, a weekly routine that persisted for fifty years. They discussed painting, wood, children, money, family, problems, peers, galleries, aesthetic movements of the past and present; the business of making art and the art of living.

"Much of the time we'd be talking about his furniture," acknowledged Harrison, "and who was the latest collector. Who was going to pay $30,000 for this or that piece. He wanted to make sure that people knew the value of his furniture. He was the best salesman I ever met."

Fellowship did not eradicate struggle. Each artist continued to grapple with his personal vision, pulling together disparate elements of craft and intuition to assemble something striking and fresh in their studios as daily life unfolded with its pressures and conventions. Harrison branched out into industrial design, fabricating tile patterns two days a week over two years for Interpace (where Millard also served as company consultant). Over ten summers, Harrison, along with his wife, Marguerite, designed crystal and ceramic dinnerware for Mikasa, first in Bavaria, then Japan, but the ceramicist always longed to return to the solitude of his own studio. Claremont proved the perfect site for a quiet life of prodigious output, free from distraction yet filled with colleagues, collectors, opportunity, appreciation.

Sam continued turning out his chairs, tables, and cradles while sculpting his woodworker's compound by tool and hand.

"He never wanted any company to try to reproduce his furniture," remembered Harrison. "Sam insisted that the only pieces going to be sold had to be made by him and his crew. He had some very good offers, some rather large companies that wanted to buy his designs so they could be mass produced. He could have made a lot more money. But he turned them down."

The two artists found room to differ. Claremont's lack of a prevailing aesthetic ideology, and the absence of scrutiny by the national press and its critical apparatus, conferred on them a spaciousness for miscellaneous pursuits. The artistic community resisted both conformity and jealousy; work was less like competition than parallel play. The former citizen-soldiers each year more emphatically assumed the role of citizen-artists.

The vignettes assembled for Harrison's one hundredth birthday exhibition speak to the enviable state of this peaceable kingdom. Harrison and Sam and Jim Hueter and Karl Benjamin and Rummy Deese, along with so many other artists, constructed a blissfully leaderless artistic enclave that could not be duplicated today without corresponding upheavals and adjustments of history to assemble in one place a like number of talented, motivated people and afford them the time and isolation to engage in the joint project of becoming themselves. They needed one another. They helped shape one another, if only by reflecting a common aspiration to make a life of art. In the process, perhaps they improved one another. (Said George Frideric Handel, Bach's contemporary, musing on his writing *The Messiah:* "I should be sorry if I only entertained them; I wish to make them better.")

Sam and Harrison proved particularly attuned to each other. Harrison, so keen on and influenced by architecture, even in his youth persuading his parents to commission a house designed by Richard Neutra. Little wonder he found himself sitting across Walter's lunch table from Sam, the master builder. And Sam, the devoted collector of midcentury ceramics, his home filled with beautiful pieces—Harrison's most numerous of all. He, too, could count himself lucky to have a friend whose work and character confirmed his own suspicions about what the good life contained.

6.

PEERS
JIM HUETER

"First of all," intoned Jim Hueter, his murmurous voice patient and measured, "I want to say that I don't know what I'm doing. I don't think any artist really knows."

It's a sentiment echoed by Henri Matisse in the catalog to Hueter's 2009 retrospective at the Claremont Museum of Art.

"In art," proclaimed Matisse, "truth and reality begin when one no longer understands what one is doing or what one knows, and when there remains an energy that is all the stronger for being constrained, controlled, and compressed."

Fortitude and reserve drive creative energy. And time. Years spent in the same place learning what's worth portraying, training one's eye to perceive the deep structure of physical space before discovering how to paint it. Since 1942, Hueter has made Claremont, California, his home. Along a gravel road close to the suburban tract developments that long ago mowed over acres of citrus groves and chaparral, he designed and built a house and studio advantageously removed from the busy sprawl. A modernist compound influenced by Frank Lloyd Wright and Richard Neutra where he could study the world's shape and color in privacy and reinterpret it with utmost feeling.

"The best word that I can hear from anybody about my work is that they were 'moved,'" he said. "What the work *is,* I don't know. But the sense of being moved is something that has to be there."

Over the decades, Hueter's paintings have wound through realism, surrealism, and abstraction, employing an array of perspectives and techniques in a questing spirit. His more recent work combines oil on wood with shellac, glass, mirrors. The pieces are beautiful and mysterious, puzzling and inviting. "Some people say, 'Oh, that work is so precise with the mirrors.' But those paintings are put together by luck. By trying and pushing and pulling so they look as if they're all finished."

Less luck, perhaps, than persistence. Hueter recalled a Halloween party a half century ago at the home of his colleague, the painter Paul Darrow. The host's son was being introduced to another guest who playfully quizzed him about Hueter.

"Is he a good painter?"

"No," the boy replied. "He isn't now. But he will be."

A smile rippled across Hueter's face, his lips patiently, almost parodically drawn. "That's what my friends thought about me at that time. I was working on it."

Mostly, he endeavored in solitude. "I've always been a loner," he admitted. "Some people would be happy working in a studio. Sue Hertel, Paul Darrow, Doug McClellan, Roger Kuntz. They were in a studio together and had a wonderful time. Joking, having a lot of fun. I'd be in another studio by myself working on these tight little still lives. But I didn't mind it."

Solitude did not signify isolation. In the midcentury beginnings of his career, Hueter could count on the company of more than a score of dedicated artists drawn to the Pomona Valley. "Milford Zornes was around," he recalled, "and Milford was a good painter, a good draftsman. And a very nice person." Frederick Hammersley, whose hard-edged style Hueter admired, taught for a decade at Pomona College. "We had something going together; we felt the same way about making paintings." The dominant presence of Millard Sheets, then assembling the art department at Scripps, provided to a great extent the impetus for Hueter to return to the Pomona Valley after being mustered out of the army. "Millard talked to me as if he had known me all my life," he recalled of their first meeting. "He was the most agreeable, most interested person." Hueter caucused regularly with painters Darrow, Kuntz, McClellan, Tony Ivins, and Karl Benjamin. "We would get together once a month. We were busy with our careers, working and showing. It was a very influential time for me. For each of us."

Later, Hueter joined what local observers referred to as "the artists' group" that regularly met for lunch at Walter's in downtown Claremont. "The group was two people," he corrected. "Sam and Harry McIntosh. Rummy Deese sometimes. I got invited somehow. Freda would ask, 'What can you all talk about for two hours?' Sam did a lot of the talking. He had good stories about the famous people

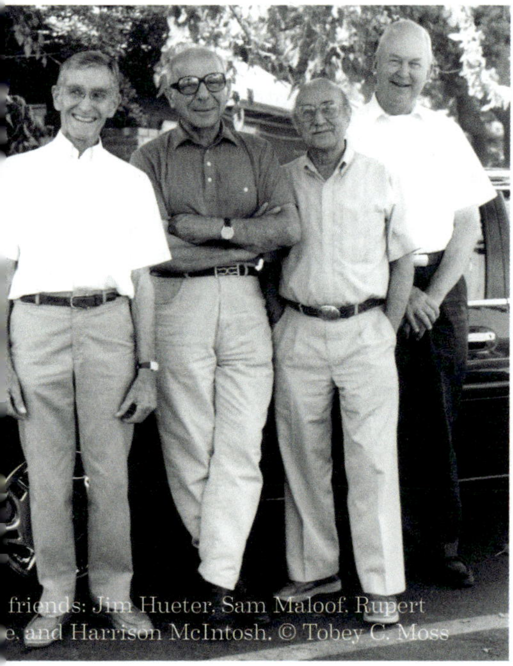
friends: Jim Hueter, Sam Maloof, Rupert e, and Harrison McIntosh. © Tobey C. Moss

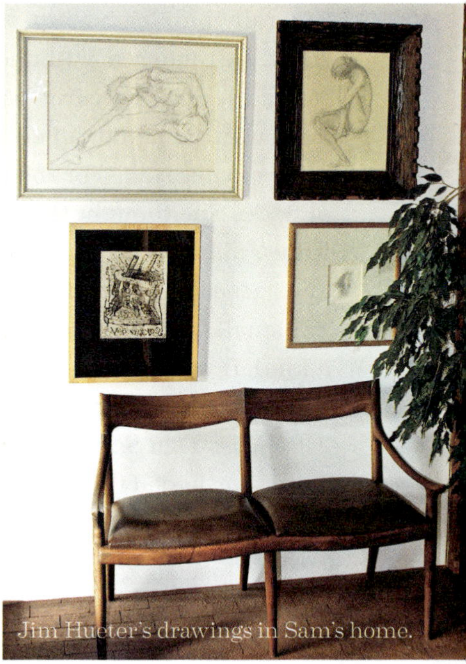
Jim Hueter's drawings in Sam's home.

he had done work for. We'd talk politics sometimes. Rummy was a very strong believer in Israel. Rummy had facts, he knew dates, names, everything. I thought, My gosh, this guy is very bright. Sam thought what had been done to the Palestinians was terrible. Sam had emotion. But they never got really mad. They just came on kind of strong at times. Then I'd kid Sam about going to the bank across the street to loan the bank money. He was getting pretty famous."

When Hueter had a show scheduled to open at the Tobey C. Moss Gallery in Los Angeles, he supplemented his paintings with an assortment of working drawings, and then started thinking about Sam. "Sam was the one who made beautiful working drawings. Why not have Sam in the show? I talked with Tobey who said, 'Why not have Harry and Rummy, too?' I was in Aptos on vacation. Tobey called and said, 'Everybody's on board. What do you think would be a good title for the show?' Right off the top of my head, I said, 'Four Friends.'"

With years and proximity allowing acquaintance to develop naturally into friendship, small favors and unremarkable services could be rendered without fuss or notice. Sam made frames for many of Hueter's early paintings. Hueter's mother purchased a large chair and end table from Sam and made a present of the furniture to her son. "For the two pieces, she paid an outrageous sum of $500. She thought that was awfully high. This was in the early fifties." Sam admired a drawing of Hueter's at a Scripps exhibition and agreed to trade it for a rocker specially designed to fit the painter's small frame. "It's an unusual one in that it's a small size. One of a kind. It was the one hundredth chair that Sam made in 1974, the last one of the year." When the Huntington opened its 2011 show *The House That Sam Built,* Hueter's beguiling oil on wood *Thin Figure Rising* joined the work of thirty-six other Pomona Valley artists associated with Sam Maloof.

Still, Jim Hueter balks at the notion that he has spent the past half century as a member of an "artists' community."

"When I think of community, I think of my close friends. And people outside of the arts who worked with and for us. Plumbers, the heater man, the air-conditioning fellow just down the block who will come anytime to help, practically for nothing. Other people too, who are very good friends." Assumptions about the gratifications of a regional artists' community mildly rile Hueter with their idealized vision of languid productivity and predictable reward "where everybody's an artist and they're all very good."

What Hueter and his peers most certainly did possess—nurtured by persistent effort and freshened over the years with innumerable conversations, questions, viewing, and use—was a standard of achievement that each artist felt obliged to pursue.

"We all respected each other's work," said Hueter. "Sometimes, I'd look at one of my paintings and say, 'Damn, that's good.' Then I would have negative feelings. I wasn't moving ahead very fast. But I started to learn that other people admired my work. People I admired who were quite successful."

Which, finally, might be the most important aspect of friendships configured to sustain a creative life. Without design or even conscious seeking, Jim Hueter found himself surrounded by people who made excellence the unspoken assumption of their efforts and tacitly expected an equal measure from their peers. It brought to the Pomona Valley a vivifying and enviable freedom.

"I was going through some drawings yesterday," said Hueter. "I had one stack I called My Little Daily Drawings. I just loved to draw and see what would happen. Just making a statement. It was pretty disciplined. I might do three or four in a row. I dated them. After a year, I had close to a thousand. One group is quite geometrical. Another, I just put some marks down in the center and scratched the head, eyes, nose. It was the most wonderful feeling that I ever had in making work because I didn't have to think about it. I'd make something, and then see what that led to."

7.

THE WOODWORKER
JIMMY CARTER

"I had a very intimate, almost a brotherly relationship with Sam Maloof," said Jimmy Carter. "There wouldn't be more than five or six other people in the world who have had a greater direct effect on my self-analysis, my deliberate effort to learn something from them and apply it to my own life."

It seems an extraordinary admission for a world statesman, Nobel laureate, our greatest ex-president.

He paced a half circle around his office at the Carter Center in Atlanta, and nodded to each of the pieces of furniture that he calls his "treasures." The ziricote low-back Sam designed to commemorate receipt of the Nobel Peace Prize. The double-rocker presented by the businessman and philanthropist John Moores. A small table delivered by Sam during one of his many visits.

"I felt there was an emotional attachment between me and Sam. There's an emotional attachment between Sam and this chair here." He paused to flash the celebrated grin. And then it vanished. "I'm sure other people had a similar relationship with Sam. To me, it was special."

During the Carter presidency, Sam resolved to fashion one of his signature rocking chairs as a token of his respect and esteem. The White House replied that it could

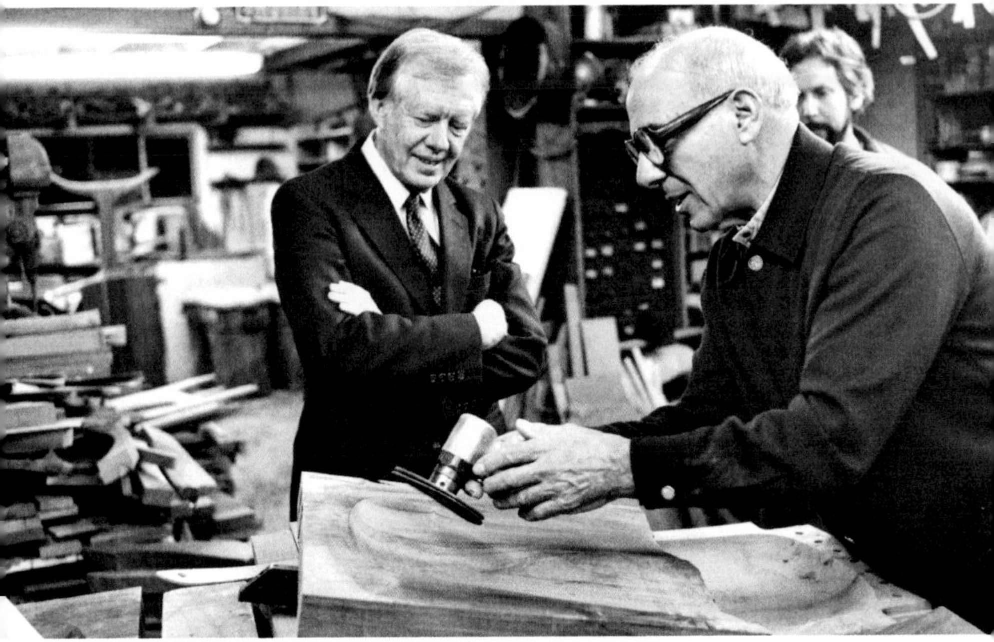

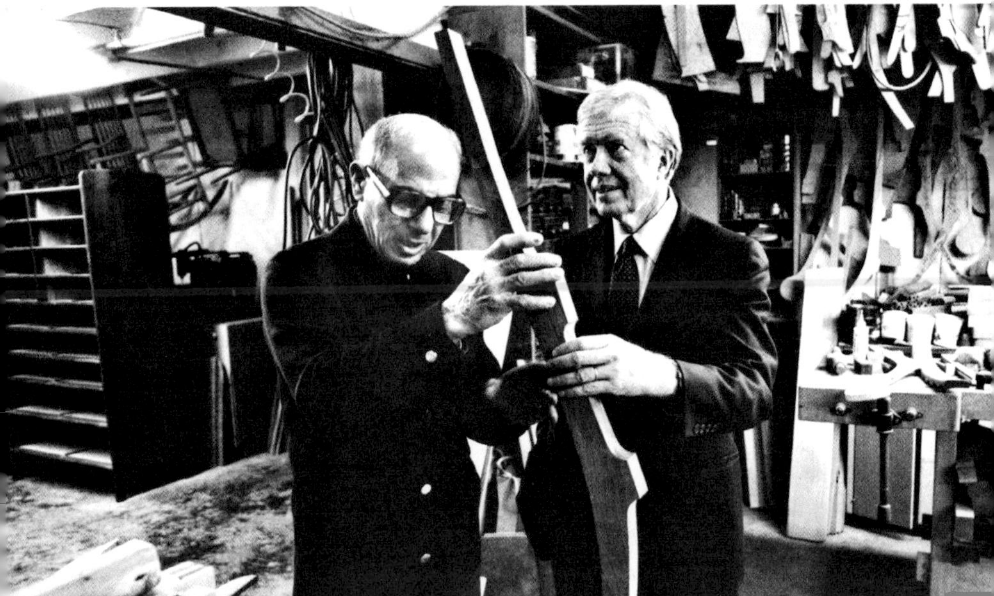

not accept gifts valued at more than one hundred dollars. Undeterred, Sam waited until the Carters left Washington in 1981. He stuffed the rocker into the back of his station wagon and hauled it 2,167 miles from Alta Loma, California, to Plains, Georgia, making the delivery in person.

Instant affinity: the two men perceived in one another some complementary glint, their alliance intense and immediate. "We spent hours and hours together at Sam's place and in Plains," remembered Carter, "and in this office." They talked frequently about the Middle East. "He was still a Lebanese at heart," stated his old friend. They discussed party politics, ethics, religion—"you can't talk about politics and Lebanon without talking about religion"—along with the Carter Center's burgeoning agenda of public health, governance, and human rights projects unfolding in more than eighty countries. "He was not just the best woodworker that ever lived," insisted President Carter. "He was a person admirable in all his characteristics. He was a philosopher. He was deeply committed to basic moral values. I really have been inspired by Sam."

Invariably, Sam and Carter talked about making furniture. Carter, a lifelong woodworker, first read about Sam in *Fine Woodworking;* he kept bound copies of the magazine on his bookshelf, and meeting Sam proved "thrilling." Of course, the same could be said for Sam, a staunch lifelong Democrat. Despite wholly dissimilar paths, they seemed primed to comprehend one another, cognizant of the rigor and tone of one another's aspirations: the interior, meditative qualities of creative labor and the expansiveness of working in the public realm. Sam often arrived bearing books and instructional videos. The president studied them and solicited advice. Sam made a present of one of his rasps—"a very high-quality wood rasp," acknowledged Carter. "Which is in my shop now."

President Carter had always esteemed manual skills; his own hands were filled with tools since childhood. As a boy, he learned to wield a saw and hammer to cut barn rafters and construct hog sheds. He graduated to blacksmithing, fitting horseshoes to mules, repairing a metal plow with a forge and anvil. "Today those things would be done on a mass-production basis. In the olden days, you didn't have the choice. You had to make them one by one, hand by hand."

Like other country boys, he joined Future Farmers of America, advancing to the full array of carpenter's tools. "Then when I graduated from Annapolis and got into the navy, I was only making $300 a month. There was a lot of difference in price between furnished and unfurnished apartments. So I would go to the woodshop on the navy base. They had master's mates who were experts at carpentry, furniture making. They had planing mills, glues. They taught me how to make intricate joints." Throughout his navy years in Virginia, Connecticut, and Hawaii, Carter stocked his home and equipped his growing family with hand-hewn furniture of his own design.

When the Carter administration vacated the White House, the staff and cabinet collected $7,600 with the intention of purchasing their former boss a Jeep. Carter confided to his secretary that he had no interest in riding around in a Jeep; he preferred to spend his days making furniture. She redirected the money to Sears & Roebuck, whose staff fully fitted out his home workshop with Craftsman tools. Woodworking, Carter later admitted, provided a welcome alternative to "thinking about having been defeated in 1980 for reelection." When he and his wife, Rosalynn, purchased a mountain cabin retreat, the retired chief executive undertook the design and execution of all their furniture.

"I've built between seventy and one hundred pieces," he estimated. Chairs, cabinets, chifforobes, tables. Toilet paper holders. Wood-turned collection plates for his church. "I built four fairly nice rocking cradles for children. To inspire my own children and some grandchildren to have babies. I've built four four-poster beds. They're all different designs, different woods. I've never built any piece the same way twice. Sam had to do that. I don't know how many of those chairs over there he's built. I have less patience than Sam had. He could build another chair identical to that one and enjoy the process. I'm not quite that...well, I'm not in the business of selling furniture."

The demands of repetition and consistency nevertheless hold a powerful appeal. For most people, the prospect of hand-cutting a hundred dovetail joints might sound onerous, but for Jimmy Carter, both as a young man and today at ninety, "that kind of thing appeals to me. It's very similar to what I used to experience plowing mules as a young man. Breaking land. We didn't have tractors or machinery. All our work was done with mules and a few horses. You'd start out on a five-acre field, plowing around and around to

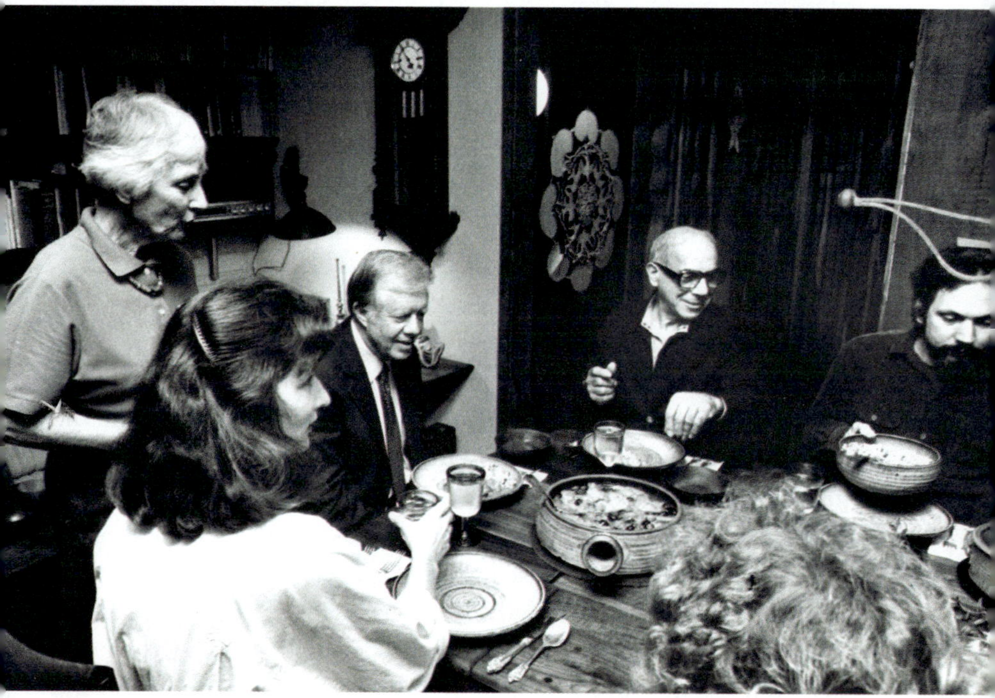

Alfreda, left, serves Jimmy Carter homemade casserole. Sharing dinner with the former president are, left to right, Kathryn, Sam, and Slimen.

break the land early in the spring or late winter. And it looks like you never will get through. Every time you make a whole round, you cover about ten inches. But when you're done, you can look back and see what you did, and you know you've accomplished the job as well as anyone could."

For a man determined to elaborate the possibilities of his life—to move from the farm to the nuclear submarine, from the statehouse to the White House, from defeat in a crucial, bitter, repudiating national election to worldwide achievements in his post-presidency—tedium proved the teacher.

"Organizing a political campaign—that's one of the most tedious things of all," admitted Carter. "I planned my presidential run from 1972. I didn't start campaigning until 1975. We did a meticulous analysis of every state. Who the influential people were, the key newspaper writers. When I ran for governor in 1970, my wife and I shook hands with six hundred thousand different people. There was not a factory shift anywhere in Georgia where we weren't there before daybreak to shake hands with the workers. All this," he stated emphatically, "is parallel to the planning of a piece of furniture. All the boring work you have to do to get it done."

Discipline implies a moral dimension: the doctrine of incremental progress. You master a small aspect of your craft over time, and with endless repetition and unavoidable frustrations and setbacks, your eventual realization of the goal (by no means assured) may provide the psychological and spiritual dexterity to undertake the next more formidable job. In the pursuit of depth and scale, we grow by micrometers.

"I'm an engineer. I like to do meticulous things. I consider each tedious moment to be constructive and part of the entire process. At the Carter Center, we quantify everything. I could tell you exactly how many latrines have been built in Ethiopia since we started. I can tell you how many people we'll treat this year, immunize, and keep from getting river blindness due to parasitic worms. Conceiving something," said Carter, "working without regard to the time or effort it takes, and then looking back on your achievement—that's how you learn how to do better the next time. That appeals to me."

His gaze fixed on the rocker, the first piece by Sam that he owned, the gift still redolent of crisp memories of its maker.

"The ability to apply self-criticism is also something that Sam Maloof epitomized. He was his own most severe critic. When Sam built something that had concealed parts, like the back end of a drawer or the underneath of a cabinet, you didn't have to worry about the backside being inferior in quality or appearance. Not one in a thousand owners would turn the piece upside down to examine the backside. You just knew when Maloof finished something, it would be perfect all the way through. It'd be the best he could do."

The president's secretary rapped on his door, eased it open and peered around the corner. There was another appointment coming up and he was already running four minutes over schedule. Once again, he flared the broad, vulnerable smile—accompanied this time by an executive's dismissing dip of the chin.

"Sam Maloof," the secretary confided moments later in the quiet, empty corridor as the president closed his door and moved on with the day. "It's his favorite subject."

8.

CHAIR
MARK PEEL

Mark Peel can picture himself as a young chef at Ma Maison, Wolfgang Puck's Los Angeles restaurant in the 1970s and the birthplace of California nouvelle cuisine. He's sitting on a case of vegetables out in the alley. At either side of him lie another two cases of carrots, turnips, and zucchini, and he's peeling and turning each vegetable in the French style—deftly, swiftly, precisely. For about four hours.

"Cooking is not what you see on TV," said Peel, a veteran of La Tour d'Argent and Le Moulin de Mougins in France, Spago in Tokyo, and Chez Panisse in Berkeley; and for twenty-three years the owner and executive chef of the Los Angeles landmark, Campanile. "When you're starting out, the chef hands you a twenty-five-pound bag of carrots and says, 'Peel these. And when you're done, I've got another bag for you.' It becomes a Zen exercise. It's the same with creating a dish. Anybody can come up with something good once. It's doing it over and over that's the trick. You have to find the sweet spot in making that dish well every time, with small tweaks and changes that improve the quality. It boils down to acquiring precision, while cultivating a dedication to perfection. Repetition and patience. Over and over and over."

It's an affinity Peel perceives in Sam Maloof's work. Patience as a dynamic virtue. The joy of repetition. The gradual abdication of bad habits and the loosening of unconscious restraints in pursuit of serene beauty and simplicity.

"Craftsmanship requires patience," said Peel, referring no less to the kitchen than the workshop. "Sam purchased wood that wasn't ready to use when he passed away. Sometimes it requires aging and takes years of waiting. Then you need to hold on for exactly the right project. Today there is still wood held in storage for Sam's shop that is waiting for the right project."

Peel paused to scour his memory for a line about craftsmanship that's often attributed to Michelangelo.

I saw the angel in the marble and carved until I set him free.

"That's Sam," he continued—whom Peel first met briefly at Campanile, and was then reintroduced to by a woodworker friend. "With wood, he simply pared away everything that was not an essential part of the rocking chair or the settee. Think about the progression of his style over the decades. You don't see any point where there are radical changes. It's a very gradual increase in sophistication. At the same time, it's becoming more simplified. That's the beauty of fine cooking, too. In great restaurants, you find the perfect item on the plate, simply done. Your fancy-schmantzy places—all those local *La Maison de la Casas*—they have so much going on that they goof it all up. And that's another reason I draw inspiration from Sam's work. I see the evidence of how hard it is, how much concentration it takes to do things simply. It takes years of practice to zero in on the essence. In the kitchen, I'm always trying to pare away the inessentials to leave clear and strong flavors and textures."

For a reminder about the power of diligence and restraint, Peel need only direct his attention to the furniture in his own home. He owns two Maloof rocking chairs, a side table, and a dining room set composed of an eleven-foot table with four leaves, two on each side, and eight chairs with upholstered feet and backs. The dining room set, equipped with a large sideboard containing nine drawers, two cabinets, and florescent under-lighting to illuminate the bottom shelf, was created in 1962 for the novelist Irving Stone—the author, fittingly, of *The Agony and the Ecstasy,* a bestseller about the life of Michelangelo.

"I got one rocking chair," Peel said, "and then a month later the other was done. My son, who wasn't ten at the time, said, 'Dad they're different.' 'No,' I told him, 'they're not. They were made at the same time from the same pattern, the same wood.' 'No,' he insisted, 'they're *different!*' So I got out a measuring tape to prove him wrong. He was right. One chair is taller and wider by three-quarters of an inch. My son saw that. When I looked at the chairs closely, I recognized that the smaller one has a tighter, more vibrant grain pattern. It's almost as if they're male and female. Even with the same pattern, each has a slightly different expression. I never asked Sam if it was conscious choice. I have a feeling he'd probably have been surprised. He always responded to the wood as much as the wood responded to him."

The aim to please the eye while serving the body has obvious significance for a chef. In his cookbook *The Food of Campanile,* Peel states that the restaurant's cooking philosophy has "very little to do with intricate plate arrangements, but is simply: does it taste wonderful and look beautiful?" To this end, Peel commends the necessity of "complete attention, understanding the rules and materials, the processes, and finally achieving balance"—all with the aim of creating a perfect meal. Yet if you fix your eye too insistently on the goal, you will lose your way. There is only one means of reaching the desired end, and that is by fully inhabiting each step along the way. Over the course of the painstaking effort, the elusive qualities of art emerge.

"You know," said Peel, "Sam never called himself an artist. Which I think is wonderful. It showed his dedication to craft. You can't achieve art without becoming a great craftsman first. You simply can't get there without the work."

Yet, the craftsman also pays a price. For Peel, the intense physical exertions of the professional kitchen—the site of unrelenting labors since he was seventeen years old—have produced a battery of injuries and ailments. Peel suffers from carpal tunnel syndrome in both wrists and shoulder pain related to thoracic outlet syndrome caused by the repetitive motions that come with endless stirring and cutting. From years of standing, he has developed a bone spur on one foot and on the other a cyst between his toes. Heavy lifting has prompted three hernia operations.

"I had a blocked artery requiring a stent," said Peel. "Did it come from the years of eating cheese or genetics? These things just add up, I think. They're occupational hazards." The one common injury he hasn't had to endure is a severe hand laceration. "I still have all the tips of my fingers," he sang out merrily. "I've cut myself badly, but nothing permanent or life-threatening."

For both the chef and the woodworker, there's an irony at play. Though celebrated for their creations in the kitchen and the shop, they both face a range of afflictions common to blue-collar life: they may be artists, but they can get hurt like working men. To Peel, this proximity to accident and injury prompts an urgent respect for the liabilities inherent to the profession, along with a reverence for its materials.

"Animals die to become food," said Peel. "To prepare it badly is disrespectful. Even plants give up their lives for our dinner plates." For the woodworker, the proposition similarly holds. "A tree is cut down in the forest to become furniture," mused Peel. "To make something of poor quality, or worse, waste it...it's all very disrespectful to the living organism."

Comingling thoughts about foodstuffs and forests comes naturally to the chef. Throughout his career, Peel has cultivated an elemental relationship with wood, though it differs in character and

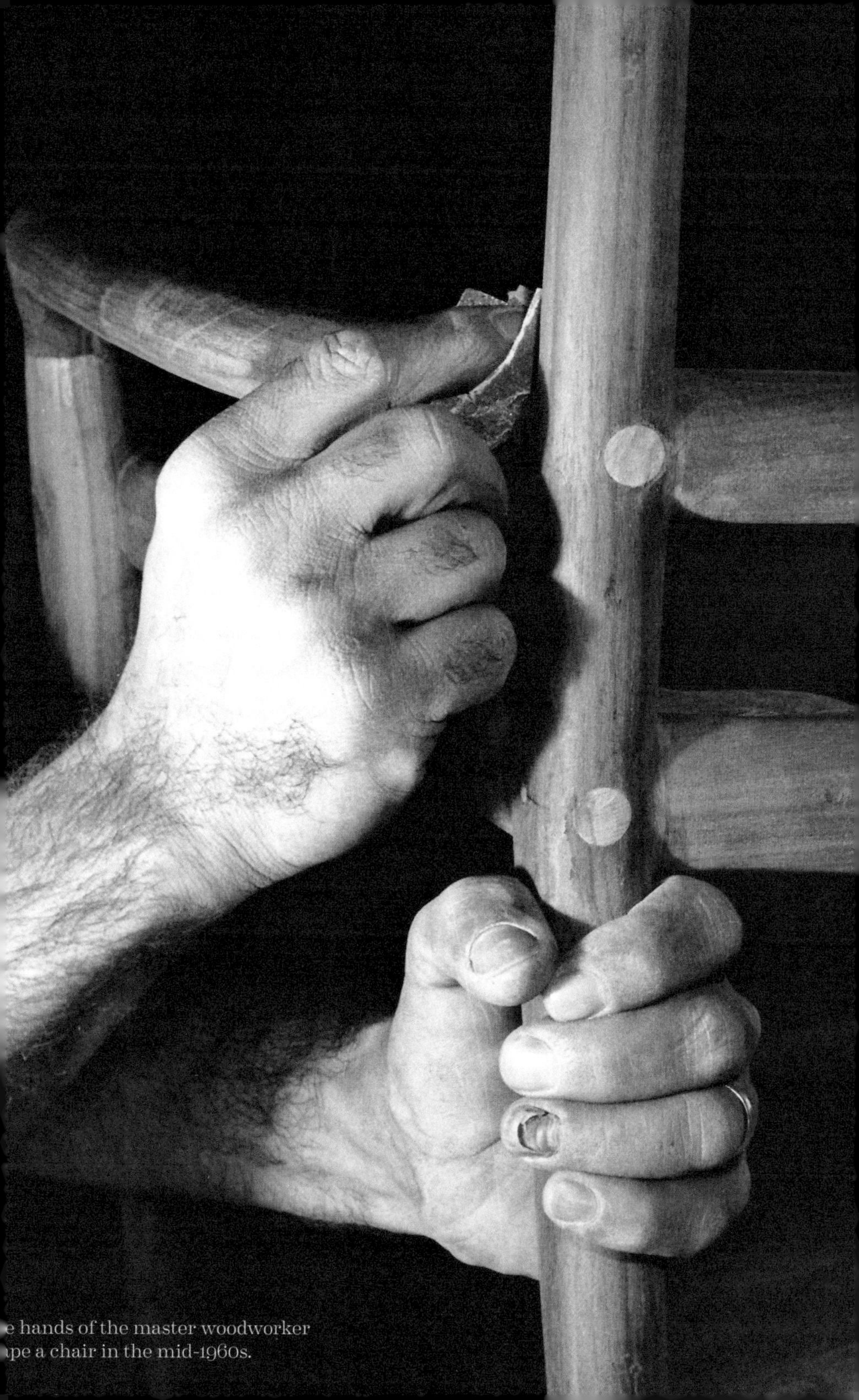

e hands of the master woodworker
ape a chair in the mid-1960s.

kind from Sam's. Jonathan Gold, restaurant critic for the *Los Angeles Times,* recently declared that "Peel is still the most exacting grill chef in the country, a master who plays his smoldering logs the way that Pinchas Zukerman does a Stradivarius." For many, Campanile's distinctive sensual expression was its whiff of wood smoke, the scent of mesquite lump charcoal and smoldering rosemary.

"You have to learn which woods are best for grilling," said Peel. "Certain woods are hot and even. Others offer up a smoke flavor, while some just aren't worth anything at all. Don't ever grill with eucalyptus. Your food will smell like Noxzema. A chef also needs to learn about the variations in heat. You can cook a rack of lamb completely on the grill, no need to roast. But you must start on a hot spot, then move to another place so it finishes slowly, so it doesn't burn on the outside and remain raw inside."

Lamb with black olive sauce and a feta cheese plate—flavors of the Levant—have been favorites for Peel to deliver to the Maloof workshop whenever he drops by for a visit. "When I go," he explained, "I bring lunch." Under Sam's guidance, lunch remained an important ritual: a collective cessation of labors in the spirit of the Renaissance workshop. A tradition not unlike the restaurant staff meal.

Peel's many visits to the workshop have also provided him with a telling insight about Sam's legacy. "The sign of a great chef," he explained, "is how many people he teaches. How many people come out of his kitchen to do wonderful work on their own."

The same, Peel believes, characterizes the legacy of a great woodworker.

"Sam's craftsmanship lives on with the people he's trained," said Peel. "Almost as great as the work he produced are the values and skills he was able to teach."

And as with the kitchen, training extends beyond technique and aesthetic judgment.

"It's as much about point of view and attitude," insisted Peel. "Sam taught precision, simplicity, dedication to craft, patience."

Peel knows from his own experience that a young apprentice can benefit immeasurably by standing shoulder to shoulder with a master.

"Very early in my career," he recalled, "I was cooking scallops at Ma Maison." Wolfgang Puck appeared at his side, regarded the dish with a sniff, and proclaimed the scallops overcooked. "'I don't think so,' I told him. Imagine, I'm arguing with Wolfgang...."

So the chef cooked a single scallop. Perfectly.

"'Taste this.'"

Rapture.

"'Now taste this.' Mine," remembered Peel. "And overcooked.... Okay, I got it. One had a chewy, funky flavor. And the other one was supple, velvety, and sweet. Right now, sitting here—I was, what, twenty-one years old?...Right now, thirty-eight years later, I can taste those two scallops." Peel paused, savoring the memory. "You have to be able to concentrate and ask yourself, What is the difference between perfection and everything else? Really focus on it. And the lesson has to come at the right point in your development so that you can truly absorb it."

Which turns Peel's thoughts back to the Maloof workshop, a vision of men at work in a way that pleases him deeply.

"What people don't appreciate is the awe-inspiring quiet precision this work requires. For the rest of my life, I'll remember being at Sam's shop and watching Larry with a piece of sandpaper in his hand going over and over and over this particular spot on a chair until he had pared it down to tenths of a millimeter. Aiming to get the curve and the feel of the arm exactly right. You can spend days on the arm, working until it's precisely the way it should be. There's no aha," stated Peel. "No reveal. No fireworks. Just patience and persistence. Until it's done."

9.

VOICES
JOE UNIS

Joe Unis recalls a rainy afternoon at Sam's house with Millard and Mary Sheets visiting, Alfreda plying the guests with tea and sandwiches, and the savory, unfurling conversation about the Pomona arts scene.

"I was just sopping it up," remembered Joe, who had been recruited in the mid-1970s by Pomona Valley Hospital to open its high-tech radiology division. He and his wife, Georgette, a painter and ceramicist, had left Washington, DC, with some trepidation about the cultural life awaiting them in a small Southern California town at desert's edge. But almost immediately, they began to meet and mix with local artists. Their friends kept mentioning Sam Maloof; he shared with Joe a common heritage as a second-generation Lebanese American. By implication, the two men should perceive in one another some atavistic, unspecified sympathy. Which turned out to be true.

The afternoon's conversation veered amiably through the history of midcentury California arts and crafts, with the two prominent artists recalling the sundry accomplishments and misadventures of their peers. Millard held forth on the design of his first house in Claremont, which, coincidentally, Joe and Georgette had purchased when they moved to town. They discussed the painter Sue Hertel, who, like Sam, had worked for Millard.

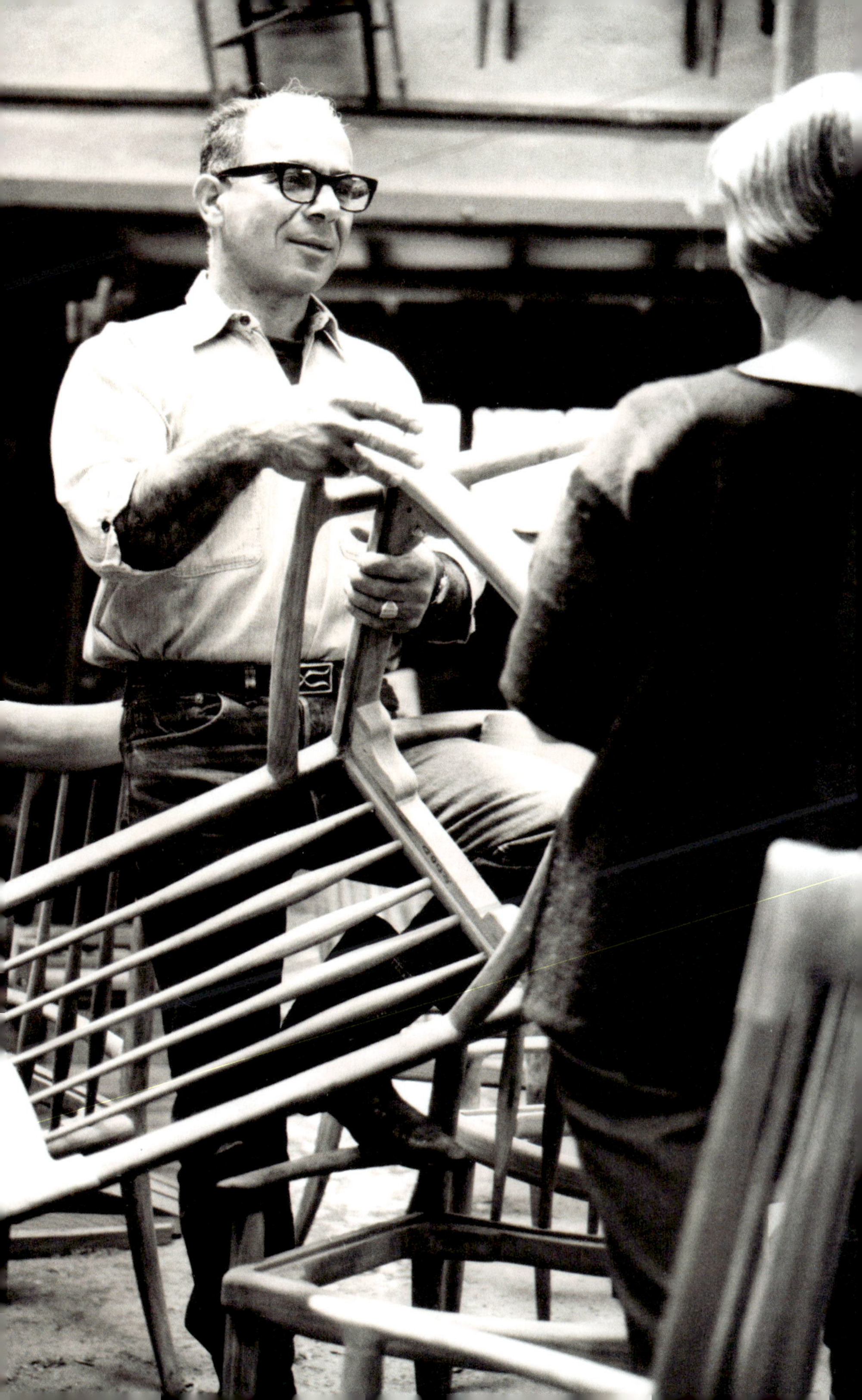

Joe knew about Sam's tenure, an exhilarating, demanding experience key to his emergence as an artist. But leaving Millard's employ had also proven difficult, contentious.

"Sam would never have said, 'Millard, you're working me to death, I've had it with you, I'm going out on my own.'" Joe shook his head and shivered at the prospect. "That would have been out of the question. Instead, he said, 'I want to give my furniture making a try.' When you're raised in a Middle Eastern culture, you don't confront. You do things indirectly. Sam was not confrontational."

Joe's own background provided other insights into Sam's psyche and habits of presentation.

"Middle Eastern hospitality," he explained. "It's welcoming people without equivocation. 'Come into my home, and let's get to know each other.' If you refuse to eat at that house, you're insulting the host. It could be seen as a repudiation of the relationship. It's hard for Westerners, even myself, to fully grasp. I had a grandmother who was a phenomenal woman, very accomplished, ran her own business. She would grumble when she was asked to drop everything and cook a meal for someone my father casually knew and had brought home. But she did. Sam expected that from Alfreda, too."

Sam had been primed by his immigrant family to presume a certain deference. "All those sisters along with the one brother who died young. There's no doubt that Sam was catered to and spoiled as a child." A possibly apocryphal and certainly unverifiable tale: When Sam was born, the bells of his family's church in Chino pealed festively to proclaim the arrival of a son. Whatever the truth, Sam entered the world blazoned with confidence. "He didn't doubt his intrinsic value. He was not afraid to engage anybody."

Of course, there's a price to be paid by the golden child. "Being from the same culture," said Joe, "I know the voices you hear growing up. 'Don't go drawing pictures. You need to make something useful.' From Sam's perspective, if he made a chair or table—

something somebody could use—he could rationalize his artistic skills. He could be at peace with the little voices that said, 'You can't be an artist, you're going to starve.' The fact that he could marry making furniture with making art was perfect."

A cunning strategy. (Has any lower-middle-class immigrant family ever exhorted their sole surviving son to risk everything on the chance of becoming an artist?) But the successful deployment of the plan also required an ideal collaborator, someone as strong as Sam, equipped with a complement of practical skills and instincts.

"If Sam had any carelessness," said Joe, "it was blind faith in human nature. Sam gave people the benefit of the doubt at every turn. Alfreda was the one who kept him from getting hurt seriously. From being taken advantage of. It happened over and over

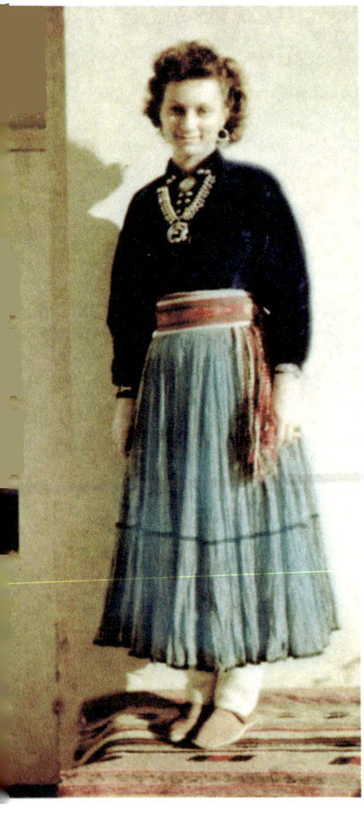

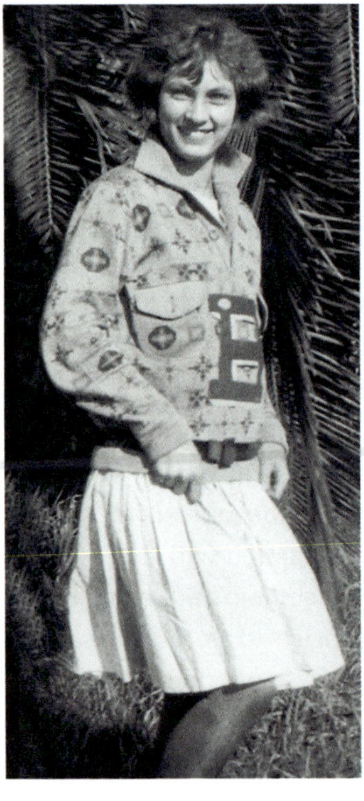

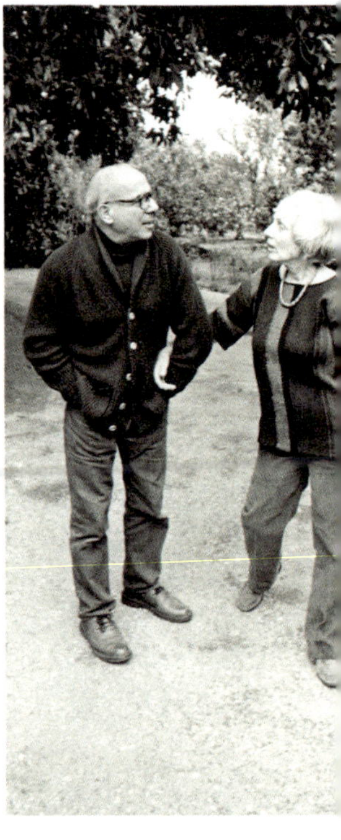

in business. Alfreda was entrusted with their survival financially. She was part of the due diligence. She made the engine run, she made everything go."

Left to his own, Sam might fail to secure a deposit for materials before launching into work. He could promise ready production when his waiting list ran into years. Alfreda balanced the books and Sam's promises. "Sam's welcoming manner was boundless," said Joe. "Alfreda was a wonderful enforcer of boundaries. She was kind, but firm. But he was only so tamable."

Committed to a profession that easily might have been conducted in solitude, Sam threw open the doors of his shop and his home, striving to extract a friendship from almost every business transaction. In the art world, few could match his charm or his intense hunger for company.

"Sam's motor was generated by other people. As long as there were people around him admiring his work, engaging him, he had incredible energy to keep going. He liked it that way; he wanted it that way. He probably wore poor Alfreda out because of his incessant need for human contact. Sam's core was really Alfreda," affirmed Joe. "He adored her, loved her, drove her crazy."

But she hardly proved defenseless. Another friend recalls an episode at home when Sam curtly demanded a glass of water. Alfreda returned and poured it over his head. "What was that for?" he sputtered. "You didn't say please. Or thank you." One senses that she was not addressing the singular occasion.

More typically, Alfreda's admirers remember her patience, the authority of her quiet voice, each syllable wielded with ardent purpose and design.

"They say as we get older," mused Joe, "that a woman will speak one hundred words for every two words spoken by a man. With Sam and Alfreda, it was the opposite. Alfreda could say one phrase. Two words. *Bingo!* And Sam would stop in his tracks."

10.

SAM'S HOUSE

Sam sometimes claimed to have a Master Plan.

His assistants knew better. At punctuated intervals, Sam pulled them off their shaping and sanding duties in the shop to take over construction of the house. Building could last for weeks, interrupting furniture production and choking off the flow of income. But money had never been Sam's chief concern. Any person working for him quickly grasped this fact, just as anybody employed at length in his shop eventually lent their hands and energies to the perpetual expansion and amendment of the house.

Oftentimes, Sam outlined his ideas for a new addition or modification after returning from one of his trips. Traveling throughout the country, he visited the homes of other builders and woodworkers who invariably regarded him as an honored guest, even a celebrity. From his close inspection of a variety of handcrafted details and elaborate architectural schemes, Sam drew inspiration for his own home, providing each enhancement with a distinctive twist. Rather than a Master Plan wrapped in a grand vision and articulated through disciplined control, his multiroom masterpiece often grew from the spirit of virtuoso reinvention. Everything that delighted Sam in the larger world was imaginatively translated into a constructive grammar that made sense for his home in Alta Loma. "The house just evolved naturally over the years," Sam acknowledged in his 1983 book, *Sam Maloof, Woodworker.* "I converted the carport into a living room, then transformed this to that and that to this, and kept adding on." The Master Plan, as such, owed everything to Sam's commitment to gradually realizing a wondrous, practical compound, a singular home that visitors would contemplate, discuss, and remember. Its individual elements—the clever door latches or sudden outcropping of gables, towers, and pyramids along the flat-line roof—could announce themselves over time.

Close observers perceived another motive behind Sam's ingenuity. He built rooms, extensions, and entire floors to domesticate the overflow of his burgeoning art collection. At times, the budding and branching house responded to newly acquired objects—the pop-out windows situating a row of ceramics in piercing sunlight, or the tree house room wrapping around the presence of a trunk. The shape-shifting spontaneity of Sam's work methods made it difficult for outsiders to follow his line of reasoning. He seldom detailed plans beyond a bare sketch; they never resembled anything like architectural drawings. In the early years, Sam assumed the primary responsibility for carpentry, wiring—whatever might be needed. Later, he reached out to his nephew Nasif, a skilled carpenter, along with Alfreda's younger family members who were experienced in construction, enlisting their aid to handle the heavier work such as framing. His shop assistants also took on the more demanding labors and then completed the insulation, dry wall, casings around windows and doorways, and all other finish work. Sam continued to make furniture, dropping by the site to check their progress. If a question arose when he was out of town, his assistants scoured the house and referred to methods used at prior stages—the project replicating itself over time. In this sense, Sam's house grew from the exertions of ambitious, inspired amateurs—

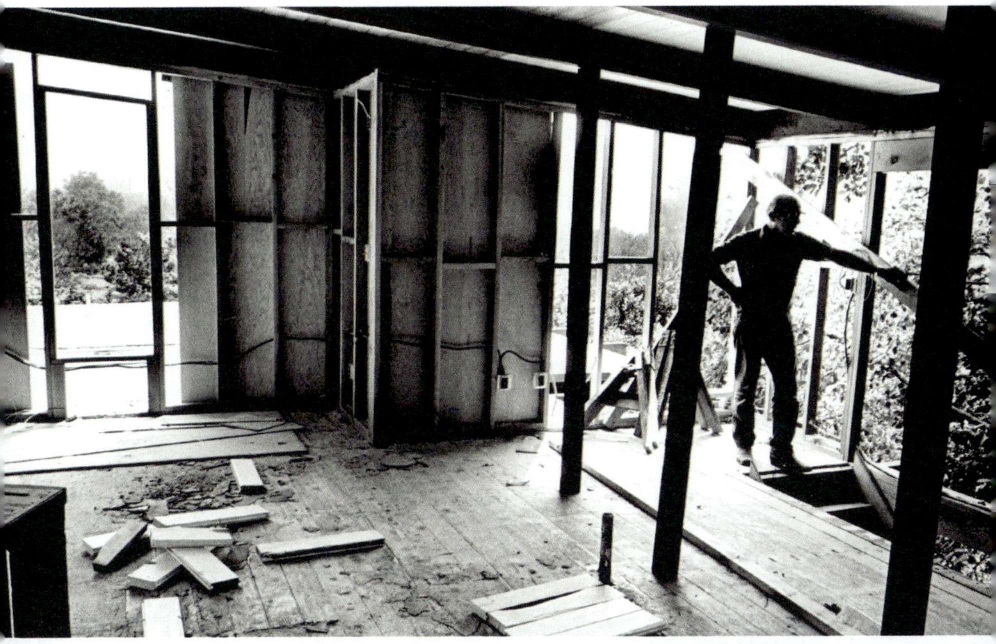

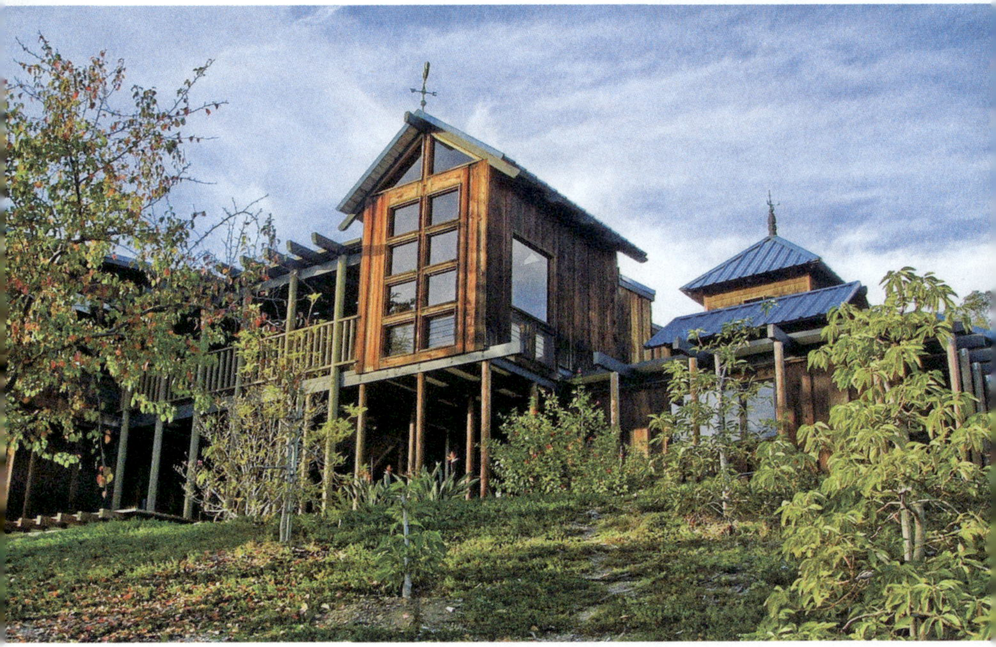

Top: Sam raised the roof to add a second floor at the house's original location.

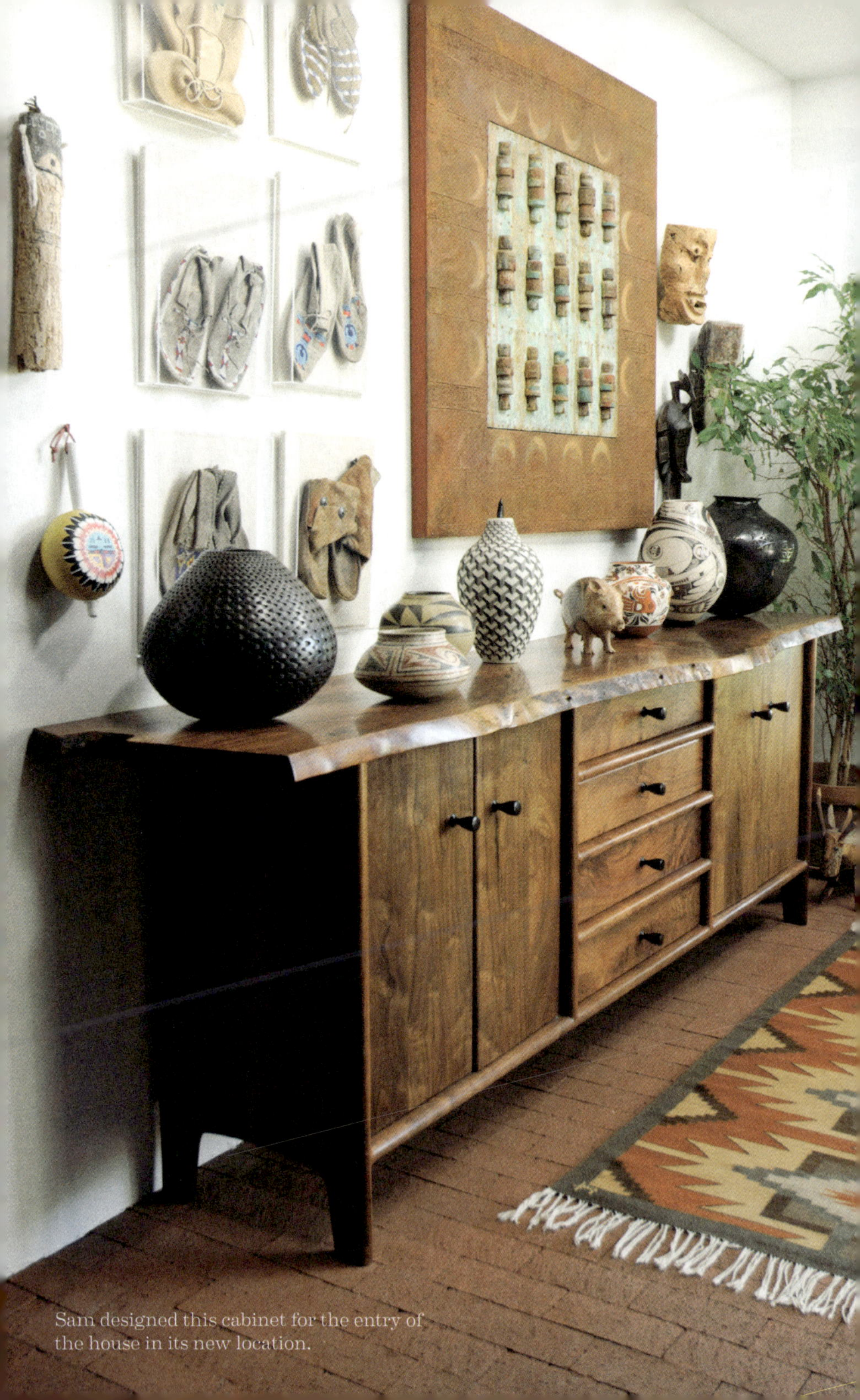

Sam designed this cabinet for the entry of the house in its new location.

nonprofessionals building for the love of the process and the gloriously unpredictable result.

In 1965, Sam and his assistants built two benches and a dining table with drop leaves for the house. They also added the first wing, comprising the master bedroom and a second bedroom. During this time, Sam's friends recognized that he was drawing immense pleasure and vitality from house design and construction. His semiprivate experiments recharged his creative energy, quelling any restlessness he might feel about repeating his well-worn furniture models for public consumption. The house offered what the furniture could and should not: a canvas for abrupt, radical transformation with the results readily viewed.

Elements of Sam's furniture found their way into house details. In the construction of hinges, latches, and cabinets, he routinely employed walnut, the preferred wood for his chairs and tables. His assistants oiled and hand rubbed wooden surfaces with Maloof finish—a mixture of linseed oil, raw tung oil, and semigloss urethane varnish—infusing each plane and rounded corner with the familiar deep-set Maloof glow. The venturesome reach of the house drew together anomalous elements and innovations: corkscrewed tree trunks employed as beams, a eucalyptus pole decoratively scarified by beetles, peeler cores from a plywood mill, and a bowed branch of Sam's beloved giant avocado tree all drawing indoors the materials and feel of the outdoors. Unabashedly sculptural features worked their way into both exterior and interior patterns: the exposed hardware of cast-iron washers, usually reserved for heavy timber bridge construction, now more ornamental than essential to weight-bearing loads.

Sam displayed exceptional nerve in tearing up his house repeatedly, negotiating the dangers and nuisance of demolition and labor. But he also worked efficiently, wrapping up modular projects with dispatch and keeping an eye on costs so that he did not, as befalls so many homeowners, build his way into financial ruin.

Not incidentally, the house served as a furniture showroom. It required comfort, clarity, informality, and order to smooth the path of Sam's presentations to customers. And if a new addition failed to meet its promise, he could always revisit, remodel. The house could patch up imperfections as it grew another layer, another limb.

Sam could curtail his building impulses so that years might pass without major changes. Yet he never stopped rearranging objects within the house, both his own furniture and his rampant art collection. New rooms and extra floors reconfigured the spatial geometry, recasting sight lines and altering the feeling tone. A walk through any room might reveal to its maker a potted plant better suited to a new portal of sunlight, a shelf of clay vessels longing to perch on the upper floor, a chair no longer aligned with the me-

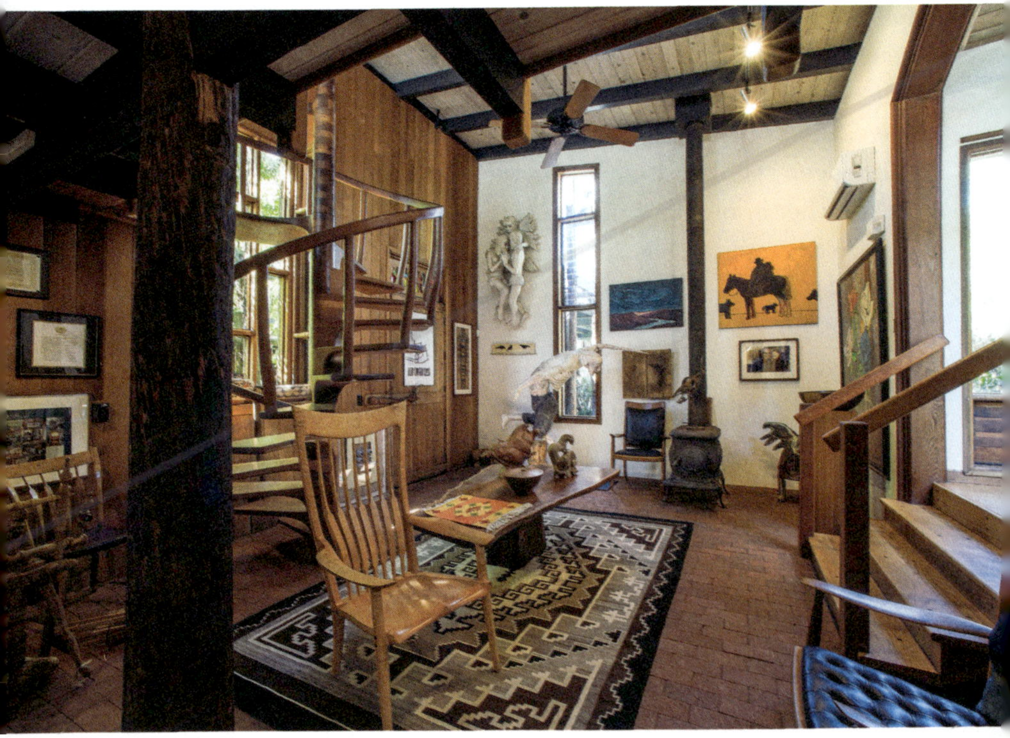

chanics of everyday use given the new doors and windows. As the Man Least Likely to Spend the Afternoon in a Rocking Chair, Sam seemed constitutionally immune to complete satisfaction. For his home, as with his furniture, there was always room for improvement. The look and feel of the house never stagnated or lost its piquant capacity to delight.

For all its interior bravado and finesse, the house offered limited curb appeal. Nestled originally in a citrus grove and flanked by huge avocado, cedar, palm, maple, and date trees, it proved difficult from a near distance to make sense out of; it was simply hard to see. When the move finally occurred, the building was sundered at the joints and hauled in eight wounded chunks under cover of night. To the midnight curious, the value and importance of Sam's creation must have seemed questionable. Once reassembled on the new property and settled upon a small mound bereft of lush landscaping, the house called to mind a storybook model of a small Western town, a dusty pioneer settlement of perplexing dimensions and untraceable origins. Some of those who had grown up with the original house—visitors and residents—expressed their disenchantment: It was not the house they had grown up with. For them, the new site suggested a kind of parallel universe—so different, yet with so much remaining the same.

Today, as the surrounding trees acquire height and the garden thrives, the house begins to recapture a variant of its original feeling. And yet Sam's house—the heart of his life and passions, his family and his career—has become something else: A museum commemorating the artist, while replicating his era and place. A staging ground for the last phase of Sam's career while he was still living, and a beautiful chute into the uncertain future now that he has gone.

11.

THE CHIEFTAIN
GARY KEENE

When Sam won the MacArthur Fellowship in 1985, he informed his pastor, Reverend Gary Keene, that he intended to increase his family's annual gift to Trinity Methodist Church. The Maloofs had long been substantial backers, first joining the Ontario congregation as newlyweds and maintaining their membership long after relocating to Alta Loma. From the start, Alfreda had insisted that their tithe be placed at the top of the family budget, an avowed priority, and Sam concurred. Faith and the succor of a faith community had always been important to Sam, though he rejected the gloomier pronouncements of the classical evangelists with their emphasis on sin and purgatorial redemption. Heaven was here and now, the world as much a paradise as we might choose to make it. A good Christian upheld his responsibility to live joyously, creatively, and generously.

Gary Keene had become pastor only the year before, moving from Michigan while still in his late twenties. Yet Sam deferred to the younger man—"respecting my role," remembered Gary, "in a way that exceeded my actual experience." He and his wife, Karen, a painter and former graduate student at the Claremont art department, promptly became close with Sam and Alfreda, deeming them "our de facto in-state grandparents." They shared meals, visited one another's homes. "We soaked up Sam's sense of style," remembered Gary. Their house décor and choice of art owed much to Sam, their social graces influenced by "the caliber of Alfreda's hospitality, which was extraordinary." Conversations between the two couples could turn intensely personal, leading Sam to introduce Gary publicly as "my other son" or "Freda's other grandson." Sometimes the closeness of their association caused Gary to question his own role. Was he pastor, confidant, or mentee? "Really," he recalled, "it was all swirled together."

Even after Gary and Karen left Southern California, the two couples retained a strong bond, often talking on the phone and visiting when possible. After Alfreda died, Gary flew out from Denver

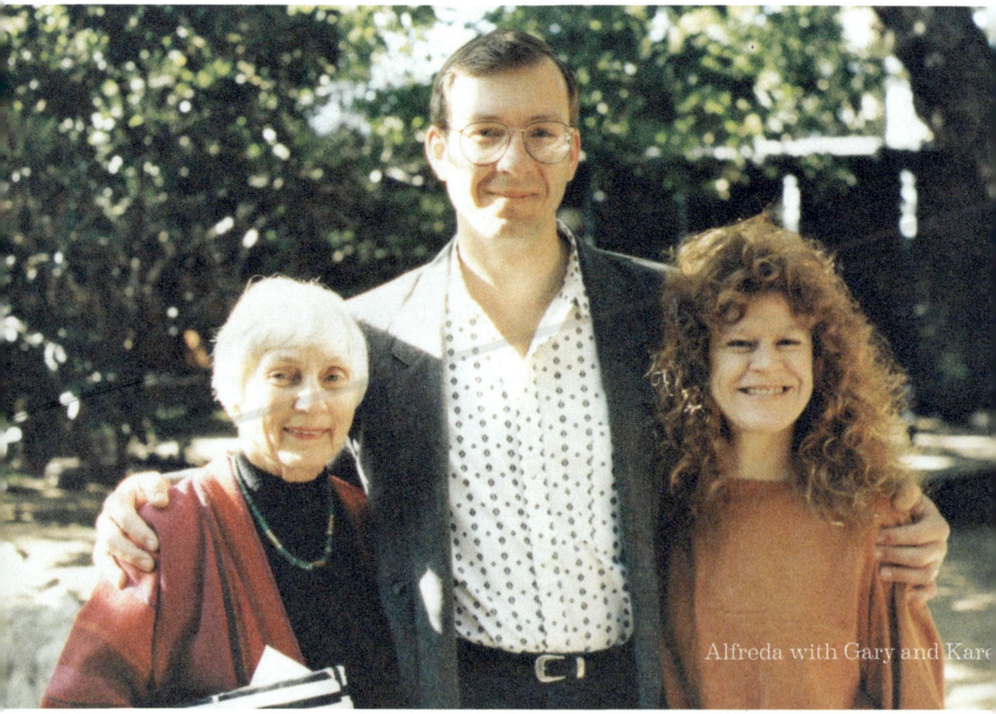

Alfreda with Gary and Kare

to spend two weeks with Sam at his house. One afternoon, they took Sam's Porsche for a prolonged ramble up into the San Gabriel Mountains. "We had lunch in a little diner. Drove down the mountain and seriously exceeded the speed limit. He enjoyed it." Around the same period, Sam and Gary dipped into the movies to catch *There's Something about Mary,* an anarchic, raunchy comedy brimming over with bad taste. "He took great delight in telling people that he'd not only seen it, but he'd seen it with his pastor. It shocked the pants off everybody. And the movie was genuinely cathartic. He laughed himself silly. It was the kind of thing he needed."

It's not difficult to imagine why Sam and Alfreda felt drawn to Gary and Karen. The Keenes were young, intelligent, highly verbal, professionally accomplished, and intensely interested in the art world Sam shepherded them through. But the younger couple also took pains to establish their intimacy with Sam on a franker, pluckier basis than convention might dictate.

"The only way to say it," recalled Gary, "is...we never took any shit from him."

Gary recognized in Sam the "Lebanese Middle Eastern tribal chieftain. He operated out of that sensibility for good and ill." On the positive side, Sam "had a very magnanimous sense of responsibility for almost anybody connected with him." He aimed to assist and shield from harm "everybody in his village. And if you met him, you were in his village." While inclined to "play the full-blown patriarchal role, he knew that wasn't best for him"—a truth often reflected in Alfreda's eyes. "Her qualities were equally sharp and distinct," said Gary. "They were like two planets that overlap, the gravity holding them both in orbit."

Sam could stumble over the encumbrances and liabilities of chiefdom, drawn off course by the expanding mass of his own reputation, and left alone after Alfreda's death. "Sam's stature and success drew a lot of people around him who were looking to get something. Very often it was financial. Or they wanted his stamp of approval—for him to open doors for their career. Or they just liked to hang around someone famous. He had a big persona. Some people need to be near someone bigger and stronger. We never gave him any of that. We were simply his friends. We showed up for dinner, and it was no big deal."

When the two couples visited galleries and museums together, Sam could dispose of his "pseudo-humble woodworker role" and speak fluently about the underlying concepts of various art movements and the course of art history. "He was very cogent. That's not a card he played out in the open very much. But he was tracking what was going on in the arts, and knew his place in it and his responsibility."

That sense of responsibility tempered an inclination toward gross egotism. "Sam was literally famous. He was on the front page of the *LA Times* when he passed. His stature and standing outside of the home poured back into the home." But while Sam took pride in his work and all he accomplished, his self-regard was counterbalanced by the weight of authentic gratitude—the source, perhaps, of the observation commonly submitted by friends and acquaintances about his "spiritual" nature. "Sam knew the caliber of his gifts," said Gary, "and he enjoyed the recognition they generated. The flip side was that he knew they were, indeed, gifts."

Another of Sam's gifts or proclivities—in fact, the aspect of his personality most frequently cited by his legions of friends and admirers—was his volubility. Sam Maloof was a great talker.

"He was going to be the center of attention, no matter what," said Gary. "He walks in and the whole room shifts to him. Once he got in the chair, he would hold that place by telling stories. And he was a good storyteller. It was his primary role in his social relationships."

But look closer at the storyteller and the chieftain, and their functions begin to conform. Although the years produced admiration and deference to Sam as master of tales, he could still find opportunity and purpose to surrender the floor—a gesture, even a tactic, that allowed him to stand back and once again observe how the world works. "There was nobody more social and attuned," said Gary. "He was always interested in people, getting a sense of them so that he could be in a better relationship with them. Those instances when he got a chance to be quiet and let somebody else carry the ball, then he could click into paying attention"—a necessity for the man with yet more stories to tell and the desire for an inexhaustible audience. And when the rare individual, or couple, withheld the customary regard and probed his assumptions with a bit more gumption and brio than he was used to, he could appreciate that too.

"That's why Karen and I became closer to him. He liked the pushback. That was the Lebanese chieftain—he liked the back and forth. If everybody was just sitting there letting him expound, it scratched his ego. But it didn't satisfy." Nor did it speak to the truth of life's brevity and the inevitable suffering that comes to every individual. "Karen and I have often lamented that for all his success, the remarkable work he did and the recognition he received, he nevertheless experienced some traumas that would have knocked out a lesser person. The loss of Freda was huge. Huge. He was never quite the same again." In the face of monumental grief, as with the years of adulation and achievement, the only apt and decent means of authentic friendship was "a raw and compassionate honesty with one another."

"We treated him as a real person," said Gary. "That's how we presented ourselves and approached them both. We're just people. We're each here with different gifts and purposes, and everybody's trying to make a go of it."

12.

ECSTASY
FATHER PETER O'REILLY

Father Peter O'Reilly had almost given up. For months, he and Sam talked about the building program at St. Maximilian Kolbe near Thousand Oaks, and the plans to construct, along with the new church, a handcrafted altar, lectern, tabernacle, ambo, and chairs. But nothing ever seemed to get started. Father Peter would give Sam another call, drive seventy-six miles across the congested 101 and 210 to Alta Loma, and wind up devoting two more hours to a leisurely lunch with Sam. Lively conversation ensued, a firestorm of possibilities. Then nothing.

"He loved to talk," said Father Peter. "He'd regale me with stories." Over time, the two men established what began to feel more like a friendship than a business association. Sam introduced Father Peter to Alfreda. "Delightful," he recalled. "And the very opposite of Sam. Very quiet, strong, and deeply in love." An active, outward person allied with someone conspicuously inward, their strengths intermingled in an unusually fruitful partnership.

But after all the meals and conversations and the evidence of their burgeoning friendship, there was still nothing that might be construed as a construction plan.

"In the process of getting down the contract with Sam," said Father Peter, "I came to understand that he had to talk things through. I'm a bit like that myself. I hate shopping on my own. If I go with somebody, I can ask, 'What do you think?' Then when I talk, I know. Sam was like that."

Father Peter remembered reading an article by a British businessman who traveled to Pakistan to negotiate a complicated deal. For days, nothing transpired except tea and conversation. Then, practically on the way to the airport, his Pakistani counterpart remarked, *By the way, about that contract...*

"I was saying, 'Can we get moving on this?' Sam was saying, 'You need to tell me your story.' The way he talked about his own life forced me to talk about my mine. So there was a collaboration. But I was not aware of it."

Finally, the long prologue concluded, work commenced, and progress was rapidly achieved.

"When Sam got going on things, he worked very, very fast. He knew exactly what he wanted to do. That was a surprise to me. But Sam's a Middle Easterner," explained Father Peter. "He had to know me first before he could work with me."

Father Peter informed the parishioners of St. Max that Sam was creating something unique, significant—"something we would hand down to the next generation. We weren't just building a structure. We were also developing an idea of who we are. We're people of the earth. We grow. I remember the rector speaking about our congregation as a great tree. We grow, rooted in the earth, into the skies. The heavens."

In an astonishing departure from contractual specifications, Sam also began work on a huge cross. "Twenty-six, twenty-seven feet high," recalled Father Peter. "I didn't know about it. I never asked for it, never thought about it. But he was going to do it."

The cross resulted from a conversation between the two men in which Father Peter introduced Sam to "The Dream of the Rood"—an Old English poem preserved in a tenth-century collection of verse, its origin perhaps centuries older. The "Rood," or rod, refers

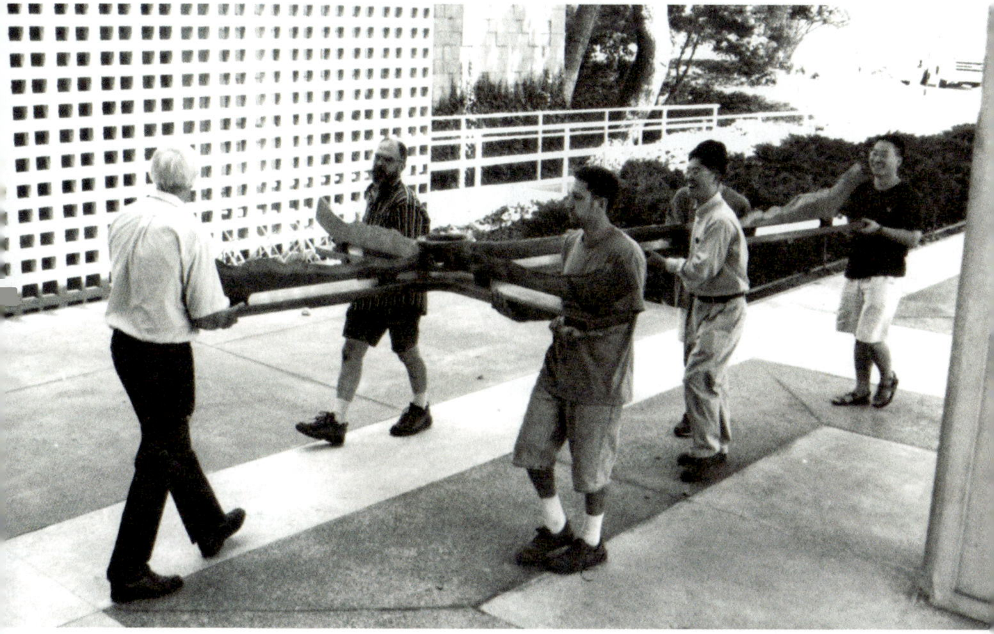

A cross for Claremont School of Theology. Several midcentury church and synagogue clients continue to take great pride in their Maloof furnishings.

more specifically to a crucifix, the crucifix central to Christianity. In the poem, the unnamed narrator describes his vision in which Jesus's crucifixion is related from the perspective of a tree snatched from its grove, cut, planked, and reshaped to form the cross. The tree speaks of its own mutilation, a counterpoint to the suffering of Christ. By the poem's conclusion, the cross and the Savior have become one, victorious in assuring salvation for humankind through their shared sacrifice.

Father Peter had first read the poem as student. He mused over its implications for years, finding himself drawn back into the declamatory rhythms of the archaic dream verse as he more closely regarded Sam's work—perceiving Sam's respect for the trees that gave themselves over to his labors, inspired by what he understood to be "the conversation" Sam silently entered into with the wood itself.

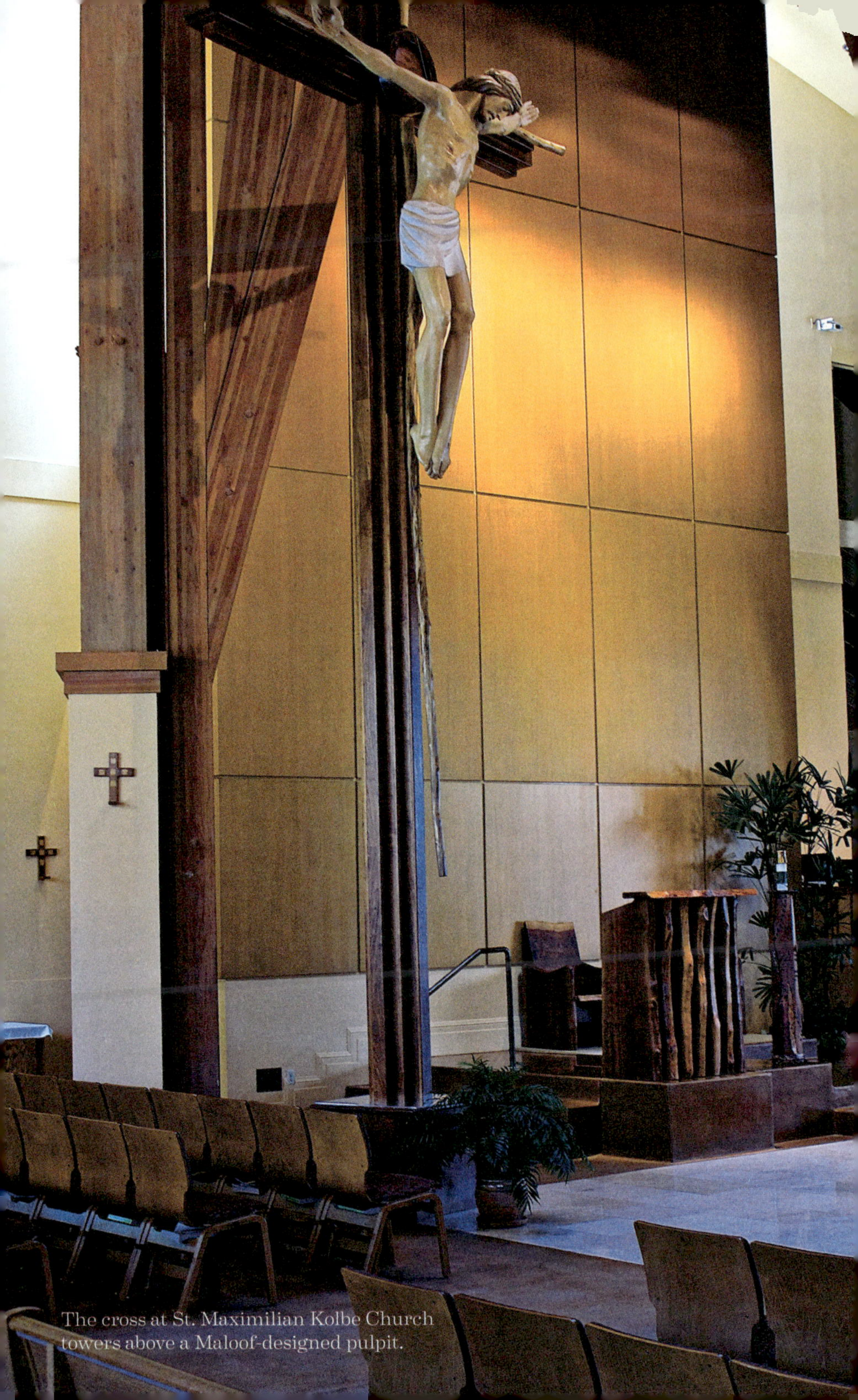

The cross at St. Maximilian Kolbe Church towers above a Maloof-designed pulpit.

As the cross—actually twenty-eight feet high, fourteen feet across—moved from construction to installation, Father Peter struggled with a conception of the new church's symbolic and visual focus that he did not originate and possibly could never share.

"Sam almost gave the cross to us as a *fait accompli*. It was so long and so huge. He had to make a special structure just so he could turn it around. There were cost factors involved too. When you're going through a building project, you're always cutting this and cutting that. Then, suddenly, Sam comes up with this idea and I thought, 'Oh, my God!'"

But Father Peter had sufficient faith in Sam as an artist that he managed his own apprehensions and suspended judgment.

"I wrote an essay one time on obedience," he remembered. "From the Latin *obedire*—to listen with attention. The mistake we make is seeing obedience as subservience. No, no. We have to listen to everybody and everything. I couldn't bring myself to say, 'Sam, thanks, but no thanks.' And I'm glad I didn't. People now go into the church and the cross is the thing they notice."

In accepting the massive, unsolicited crucifix, Father Peter had surrendered to Sam's intuition, his own vision of the Rood growing into the heart of the Thousand Oaks parish. As the crucifix became seasoned with time and use, Father Peter perceived a rightness in its presence residing below words and reason.

"The mind is like a mill," said Father Peter, "that grinding on and on. We're very much into thinking, thinking, thinking. It gets in the way. Maybe we need to evacuate all that and let ourselves be filled with silence. Then you're in ecstasy. Which means to step outside yourself. Ecstasy is allowing ourselves to be astonished. Something we stand back from, don't control, letting it speak to us. I'm speaking about awe."

13.

FAME
SLIMEN MALOOF

"Thinking about my parents," said Slimen Maloof, nibbling his way through the lamb stew lunch special at Walter's restaurant on a warm spring afternoon, "that was a really ballsy thing to do when they left that little tract house in Ontario. Alta Loma was like the backwoods at that time."

An eccentric choice, out of step with California in 1952. Waves of nondescript subdivisions were already sweeping south, eventually to pave over the citrus groves and grape vines that ran up to the San Gabriel Mountains. But Sam and Alfreda nurtured a vision for their young family: they would leave suburbia to live pleasingly out of step with their surroundings. Sam began building a home within a grove in the midst of a small town adjacent to the suburban onslaught in the fastest-growing state in the country.

A congenial setting for Slimen, known then as Sammy, frequently under pressure from throat infections, chronic tonsillitis. "My tonsils were so swollen," he recalls, "they couldn't remove them. Finally, I had to have this tracheotomy." He sets his fork on the table and fingers a scar on his throat. "For two or three years, I was really screwed up. Almost died. Those first years, I was not all there."

As he grew older and stronger, young Slimen recognized the wisdom of his parents' relocation. He rode his bike everywhere, staged orange fights at the edge of the groves with his pals. "You knew every person in town, how many kids they had, everything. I had a very normal childhood," he remembers. "It was never like Sam Maloof was anything special. He wasn't on a big pedestal."

The pre-famous Sam. He got up each morning and went to work. Slimen, unlike most sons, could watch his father on the job, his shop inside a converted garage a few yards from their house.

"Over the last few decades, everything Sam made was shipped out in wood boxes around the country. In the beginning, as a little kid, I remember we'd drive into LA and go to Bel Air and Brentwood to deliver his furniture. Most of his clients were very nice. They would offer him food or a Coke. But he wasn't treated like a rock star. It wasn't like they were buying from some famous artist. I remember going in the servant entrances. He was almost like a manual laborer. Maybe a couple of steps above the gardener. I hate to put it in these terms. But there was definitely a class issue back then."

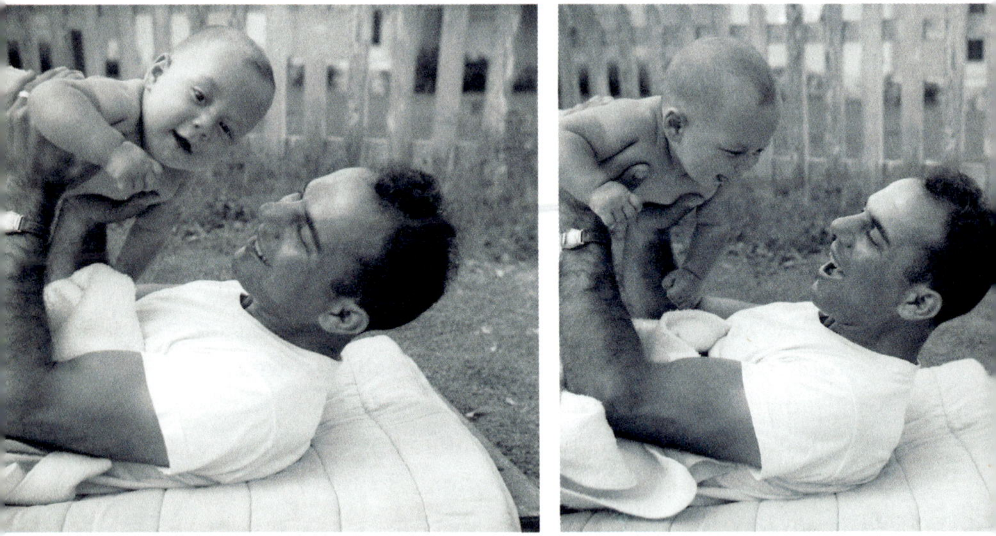

Baby Slimen with his dad.

In the 1950s, craftsmen did not enjoy celebrity status, nor were they regarded as rightful kin to painters and sculptors. Sam obviously demonstrated gifts and abundant industry, but the world does not want for talent.

"It's unbelievable how many good woodworkers there are," agreed Slimen. "Still, most can't design and they can't shape. They don't have the patience to take the piece to where it needs to go in the shop."

Throughout Slimen's childhood, Sam gradually altered his designs in ways that increased their appeal, while he patiently mastered the art of shaping. But the task of making a living still very much occupied Sam's daily concerns. Years passed before the shop did better than break even.

"Sam was giving the stuff away," said Slimen, noting the low cost of his early pieces. "For as long as I could remember, he had a huge backlog of orders." Slimen argued for a more realistic pricing scheme. "'Obviously, you're selling the stuff too cheap,' I told him. 'You need to raise your prices and have two years of backlog instead of three to five.' Sam's argument was, 'That's money in the bank, that's my security blanket, you don't know what the hell you're talking about.' I was just a kid, a rebel, a hippie. My mother was the one who raised the prices when she saw what fly-by-night woodworkers were getting for their stuff. Sam was a hell of a worker," emphasized Slimen. "A hell of a self-promoter. And people loved him. But he got taken advantage of a lot. In his early days, for sure."

The MacArthur Fellowship in 1985 brought about changes.

"They had very little savings when he received the award," Slimen recalled. "Then they got $375,000. It made them breathe easier, gave them a bit of a nest egg. It helped with his recognition. And I think it kicked my mom to raise the prices."

More invigorating to Sam than the money was "the recognition, the publicity. He loved going out into the world and talking about himself. It came natural." Sam delighted in his standing among a cadre of certified "geniuses"; the MacArthur's roster during his year included choreographer Merce Cunningham, poet John Ashbery, literary critic Harold Bloom, cultural anthropologist Jared Diamond, activist and Children's Defense Fund founder Marian Wright Edelman, and Shing-Tung Yau, a mathematician known for his success in "applying differential geometric techniques, especially the theory of minimal surfaces, to areas as diverse as general relativity, representation theory, and algebraic topology."

"Yeah. He just ate it up."

With the MacArthur, the class issue reasserted itself, but on different terms. "The tables turned," said Slimen. "When Sam got the MacArthur, people started crawling out of the woodwork." The business now attracted conspicuously affluent customers, individuals prominent in their own fields who perceived in Sam's eminence a suitable consanguinity. As a MacArthur fellow—and the only craftsperson at that time to achieve this distinction—Sam's status accrued outside the realm of mere wealth. Genius trumped cash. Exuding charm and vigor, while offering his friendship with an unaffected ease, Sam's attractions to the wider public proved irresistible.

"What is fame?" asked Slimen. "There are so many levels. Being famous as a woodworker is a lot different from being on the cover of *People* magazine, like a Kardashian." Sam invested his fame in furthering his work, arousing interest in his house and furniture while cultivating his circle of collectors and knowledgeable enthusiasts. "That's one thing I'll say about my dad—he was a down-to-earth guy. A happy-go-lucky guy generally."

To be the son of a famous man is a complicated proposition. Fathers start out as gods to their children, their status gradually diminished by the corrosive of time and the unavoidable tendency to respond to everyday challenges as a fallible human being. To find the ordinary man who is your father suddenly regaled by strangers as an incomparable figure, deferred to with a kind of public reverence normally reserved for the intimacies of childhood—it can feel off-kilter, invasive, sometimes a bit absurd. (Can anyone make sense of *People* magazine's alliterative quip to its 46.6 million readers proclaiming Sam "a Hemingway in hardwood"?)

Slimen remembers the uncelebrated Sam. "Sam was a good dad; he was involved. He did some cool things. Stood up for me." Slimen recalled a fight one day after school when he was in sixth or seventh grade down on Monte Vista Street in Alta Loma, the neighborhood called Shack Alley.

"My dad was driving up the street, Amethyst. Monte Vista butted into Amethyst. He sees me and some buddies out of the corner of his eye, down a few houses. There's this huge crowd of people." A friend of Slimen's had squared off against a hulking adversary nicknamed Caveman. "He was about twenty years old and still in, like, eighth grade." All these kids in the street, and even some adults, everyone baying for blood. Slimen's friend had already taken a bad beating. He scooped up a weapon—a large rock. His adversary picked up a four-by-four. "He probably would have killed my friend. He was a big kid, but Caveman was a caveman. Didn't get that name for nothing. Right then, my dad drives up. Runs out there into the street. Just starts yelling for them to stop. Tells all the parents they ought to be ashamed of themselves. Tells me to get in the car. Normally, you'd be really embarrassed about that. But man, I was relieved. Because I swear to God, my buddy would have died right there on Monte Vista. To this day, he agrees. 'Yeah, Sam probably saved my life.'"

14.

FELLOWSHIP
JENNIE CHINN

Their car would pull onto Whittier Boulevard, reach the edge of town, veer left, and then the city would disappear.

"We'd be in the country all the way to Sam's," remembered Jennie Chinn. The 1950s: no spiderweb of freeways, hardly any traffic. Carving a pathway through the citrus groves to Sam and Alfreda's house felt like an exploit, a thirty-seven-mile adventure. And then the minute they arrived, "this joyous, happy man" plunged out of his front doorway to greet them.

"Harlan! Lydia!" he'd shout. "Jennie Ann!" Sam always coupled her first and middle names. Nobody else did that.

"It seemed that every time we'd go out there something new was happening," Jennie recalled. "Another table or chair materializing in the workshop. An addition to the house. A cabinet latch ingeniously refined—and everywhere the smell of linseed oil. "I remember him saying, 'Jennie Ann, come see what I built.' Sam pointed out every detail of a door handle and explained why he did what he was doing."

Jennie's father, Harlan, had shared an apartment with Sam in the 1940s while they were both working at Angelus Pacific in Los Angeles as graphic designers—"commercial artists" in the argot of the day. Graphic design paid the rent,

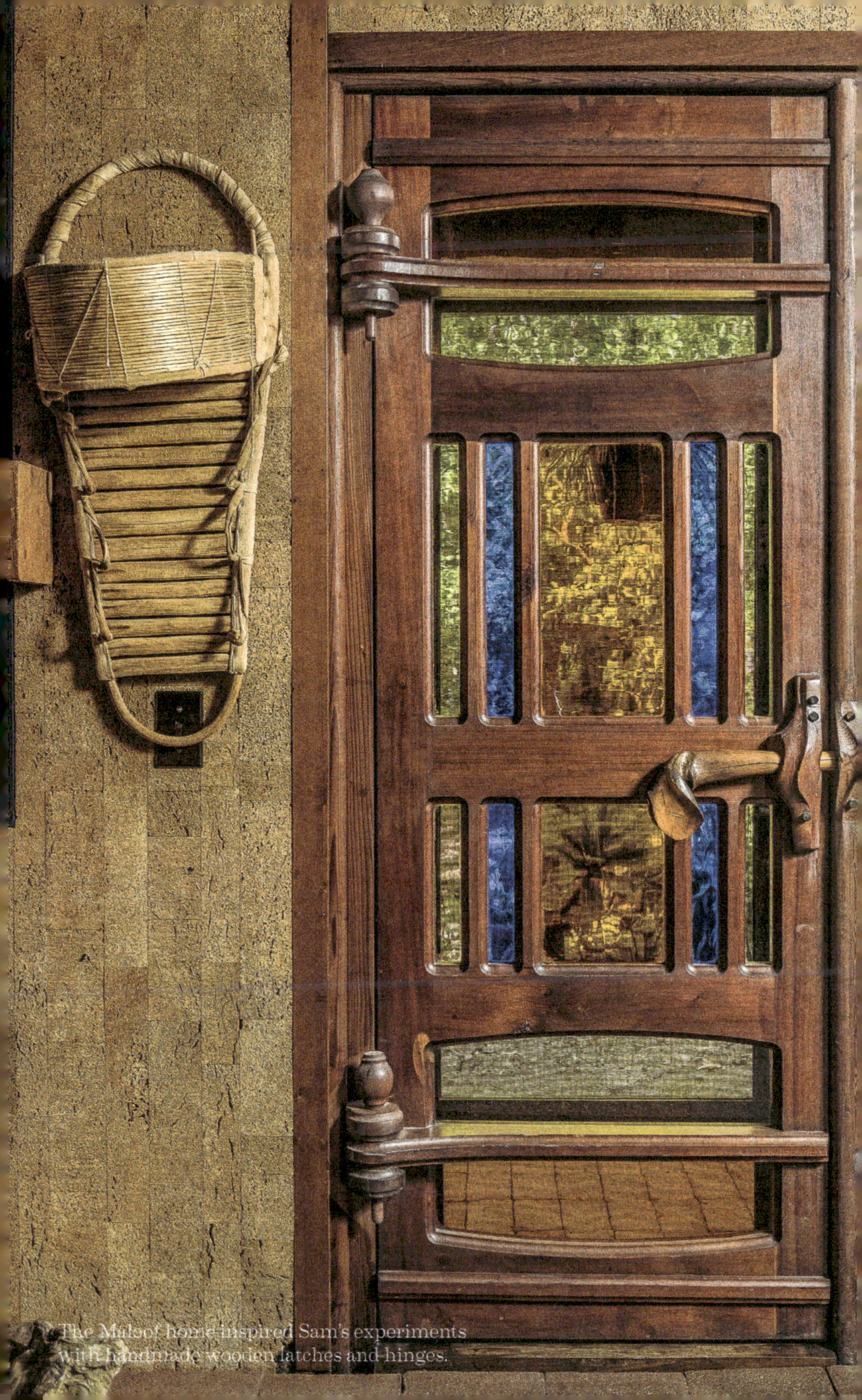

The Maloof home inspired Sam's experiments with handmade wooden latches and hinges.

but both men lived for art. Harlan had attended the Chouinard Art Institute before the draft shuffled him off to war at the age of thirty-six. When he returned home and started a family, he elected to take a day job, reserving the evenings and weekends for painting watercolors, then making ceramics. Books filled their modest house in Whittier, including an extensive collection of color-plate art books. "My parents may not have gone to college, but they were as educated as anybody I've met. As were Sam and Alfreda. Part of it seemed to be the way they all just drank up knowledge. It was a wonderful time in Southern California. It wasn't crowded. You had opportunities, you had nature. All these people moving in after the war who saw the possibilities. A belief that they could build this good life, it could be what they wanted it to be."

Their artists' circle expanded, friends of Harlan's quickly becoming friends of Sam's and vice-versa. "Our extended family," Jennie acknowledged. Their work began to show up in Claremont exhibitions, Orange and San Bernardino County galleries, the Los Angeles County Fair. "Sam's chair, maybe some of my dad's pots would appear together. At the time, I took Sam's furniture for granted. Everybody made something. And these artists were all very social. The dinner parties....I remember the food in handmade pots, all of us sitting on handmade chairs. They would celebrate their art by using it together. It seemed a form of communication. Some people might use the word *fellowship.* In those early days, they didn't seem to need the money. The art was the reward in itself."

Jennie's mother wove and tailored all of her daughter's clothes. Her father sometimes silkscreened patterns on the fabrics. "At some point in sixth grade, I wanted a dress from Penny's. But the handmade was always valued over store-bought manufactured goods in my childhood. When I got to college, I realized what a gift that was. What a magical upbringing I had being around these highly creative people."

From her parents most decisively—but with Sam, Alfreda, and a confederacy of artists in adamant support—Jennie received an enviable legacy: an aesthetic view of the world. Beauty resided everywhere. Art could be functional. She recalled the conversations at her house and Sam's about American Indian art. Alfreda relating stories about teaching at the Indian School in New Mexico. A bracelet from the Navajo reservation that Alfreda gave her. "At the time, few people were as interested in indigenous art forms. They all had a great appreciation." Art, Jennie realized, could be created by "ordinary" people. Evidently, by one's own parents and their friends.

"I was fascinated by the stories they all told. At our house, I would try to stay up as late as I could to listen until they ran me out of the room. Which is interesting, given the work I later went into."

As a graduate student, Jennie studied folk art, serving as a librarian at the Folklore Center at UCLA. In Kansas, where she relocated in 1980, she conducted oral histories as the state folklorist and later became executive director of the Kansas State Historical Society.

"I never stopped being a folklorist. That's where my heart is. It's totally connected to my upbringing. My parents and their friends took scraps and things they found and made something beautiful out of them. They taught me without telling me that all human endeavor is valuable." As a result: "I'm very interested in everyone's story."

In a feat of poetic coincidence, Jennie also serves as the Kansas State Historic Preservation Officer—a position whose counterpart in California assisted the relocation of Sam's house once freeway planners had confirmed the route through his property. "I remember the anxiety as a teenager about the freeway coming through. 'Oh, my goodness, they're going to take the house.' That was a topic of conversation because it seemed to go on forever."

Yet the dominant tone of those years wasn't anxiety, but friendship fully reciprocated. When Harlan aimed to make some renovations around his family's house, Sam lent a hand. "It wasn't ever a one-way street with Sam. It wasn't like there was a single star in this relationship." Sam appreciated Harlan's ceramics and Lydia's weaving; they worked out exchanges of their work. When Jennie's local theater company staged *Barefoot in the Park,* her father volunteered to build the sets and recruited Sam and his shop to finish them.

Jennie recalls a moment in her teens, following the two men into one of the rooms in Sam's house where they began flipping through a stack of Navajo rugs, discussing the craftsmanship and design, the colors and materials. Then the conversation pivoted. Sam made some earnest, now forgotten gesture and the two men began to speak urgently about how much their friendship had meant to each of them over the years. "That stuck with me my whole life. It was a surprising conversation for me to hear."

Yet it fit with what Jennie had observed since childhood. Sam or Harlan might pick up a piece of art—a pot, or a small painting, or a ropey plate of textile—and they'd study the object, discussing what they perceived. "At Sam's, they would be caressing the wood," remembered Jennie, "talking about what wood could be. At our house, they'd be caressing the clay and talking about what it could be. There was such gentleness and respect for what they were holding." Yet their words meant less than the action itself, the impression that they were grasping the world in their hands. Saying, in effect: This is it, this shard of human longing and effort, this item composed to delight us. This is what matters more than we can say.

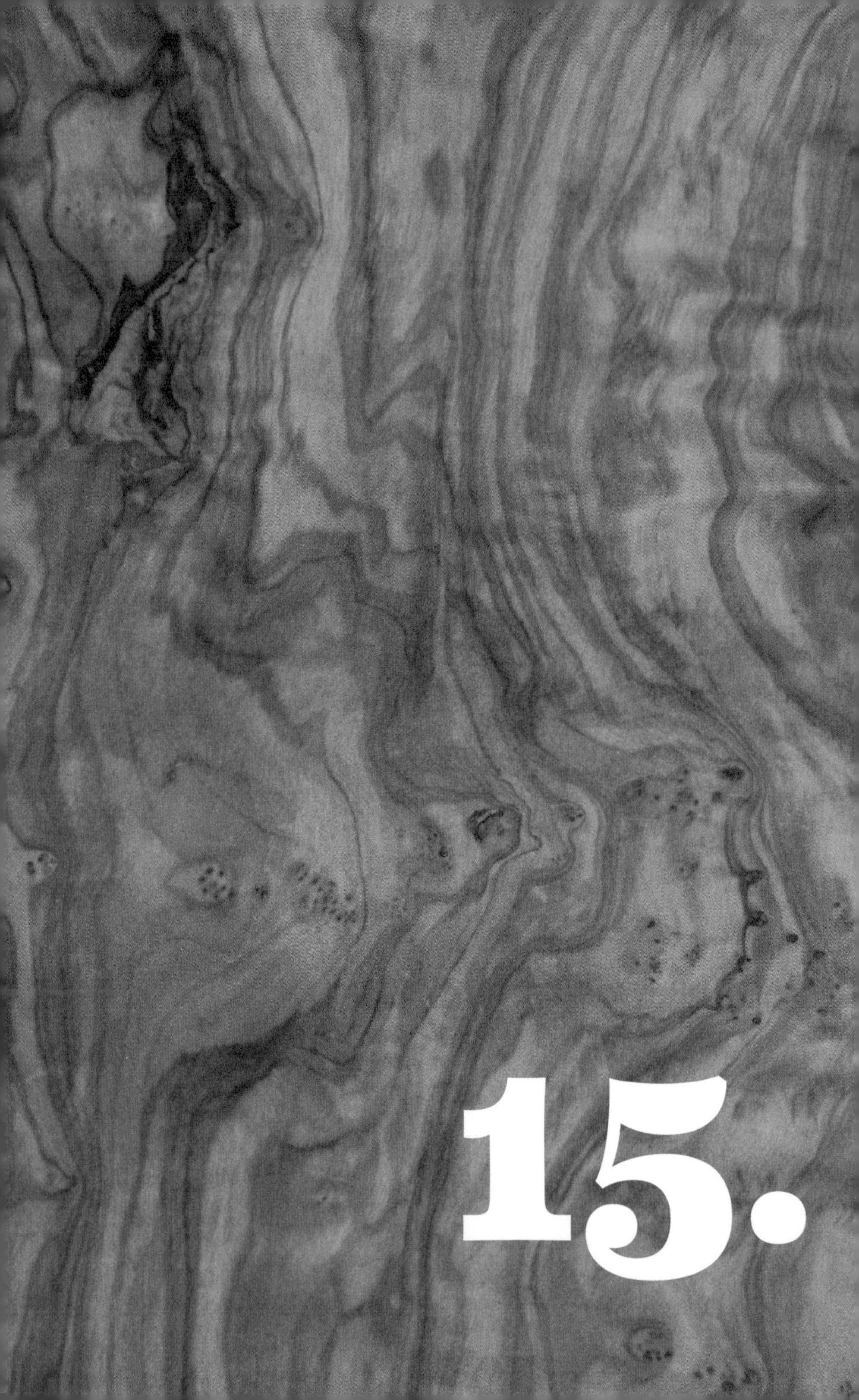

15.

VISION
CATHERINE MCINTOSH

"I guess I just took it for granted," said Catherine McIntosh, daughter of the ceramicist Harrison McIntosh. "You make things and people buy them."

In elementary school, Catherine invented a cartoon character she named Bug Me. Early 1960s: era of the cardboard Wacky Plak caricatures accompanying stiff, pink sheets of bubblegum sold at the new supermarkets, and the ubiquitous Rat Fink, a grotesque eyeball-bulging inversion of Mickey Mouse created by Ed "Big Daddy" Roth, the Southern California hot rod detailer. Catherine's Bug Me found its market among her fellow fourth and fifth graders on the school bus. She designed a set of thirty-two colored pencil drawings on three-by-five-inch paper—Bug Me on a surfboard, Bug Me on a skateboard, Bug Me with a lion. "The kids would save their pennies and I'd take their orders. 'I want one of number seven and number twelve.' I was an only child and we didn't have a television. At home in the evening, it gave me something to do. This seemed totally normal. Dad made pots and sold them, and I made drawings and sold them."

Catherine moved on to crafting pendants, her technique abetted by her father as she wandered through his studio after school. "I made butterflies and flowers and peace signs, of course." Local shops carrying Harrison's ceramics included some of his daughter's work. "I was plugged into Dad's market." When Catherine entered eighth grade, somebody ordered a gross to hawk at the Chicago Gift Show.

A childhood spent watching her father work at his potter's wheel and years of gallery-going with both parents cultivated an unusually strong eye. "Dad would keep up with the art world by reading magazines, the *LA Times*—he knew when there were significant shows. I was taken to museums when I was three, four, five years old. They didn't leave me at home with a babysitter. It was too expensive." Throughout the sixties, family expeditions to the museums and galleries in Los Angeles and Pasadena turned Catherine into "a very good looker. Ask me what was sitting on the table years ago, and it's like pulling up a color photograph from my brain. I can remember colors precisely, but not five notes of music or a phone number. Recalling that an old pillow was a particular shade of brown and green—that's no problem."

If there's any semblance of an art gene coiled somewhere in human DNA, then Catherine McIntosh seemed a likely candidate to express it. Her father and his brother had devoted their own youth to exploring the Haggin Museum in Stockton, their experience sorting them respectively into ceramics and painting. Catherine's mother, Marguerite, "grew up going to the Louvre ever since she was little—that French sensibility." Her mother's father, Marcel Loyau, sculpted huge monuments across France, though he died at the age of forty. (His eight thirty-foot bronze sea horses can be viewed at the Buckingham Fountain in the center of Chicago's Grant Park.) Catherine's grandmother Jeanne Morancé painted in the Impressionist style; her own father had been an academic painter.

Of course, it's culture, opportunity, parental influence, and preparation that coax along whatever intrinsic interest and ability might reside in the deeper recesses. Catherine took her place at the family's Maloof table—"it sat six, maybe seven people, if you squeezed"—as her parents filled out their dinner parties with the painters, sculptors, and woodworkers then assembling the Pomona Valley arts scene. Claremont proved the ideal small college town for a young girl to gradually enlarge her compass of visual enthusiasms and perceive the practice of art as a viable profession. The sculptor Betty Davenport lived down the street; Catherine played with her daughters. Sculptor Albert Stewart lived across the street, though he passed away when Catherine was still in elementary school. Painter Milford Zornes also owned a home in the neighborhood, and Catherine knew the painters Jean and Arthur Ames "pretty well" as a child.

"I used to take care of people's cats and dogs when they went way on trips. Bring their mail in. I was in a lot of people's houses up and down the block. Swam in everybody's pool."

Her own family home had been designed by one of Claremont's leading contemporary architects, Fred McDowell, as was the home of painter Karl Benjamin. Even as builders stamped out the Southern California suburbs in numbingly rote precision, Catherine's eyes fixed on the clean lines and smooth surfaces of Richard Neutra's modernist innovations. Neutra designed her grandparents' nine-hundred-square-foot house (two small bedrooms, a minute bathroom, a hallway-sized kitchen) and another friend's residence nearby.

"To me, they were normal houses. If anything, the tract homes were unfamiliar. Those little boxes. It was like, Why would people choose to live that way?"

For her parents and friends, Claremont felt "like their little Shangri-la. They had the beautiful mountains, the town was away from things, the weather was easy. You had a lot of culture in Claremont and you could easily get to LA or Pasadena." As she entered her teens, Catherine grew increasingly absorbed in matters of shape, line, color, shadow. Still, she felt no compulsion to follow her father's path into ceramics. "Every pot I tried to make ended up looking like a bad McIntosh pot. I needed to work in a different area." She had begun taking photographs as a child and continued through high school and college, preparing herself for her future career in newspaper journalism as a photo editor. "As a teenager, I had the big silver case with the long lens, the wedding lens, all that." She was earning her seat at the Maloof dinner table full of artists in a town profuse with all manner of arts and crafts.

A snapshot called up from memory: Sam leading her family over every inch of the kitchen soon after he finished another corner of the house, a perpetual work in progress.

Another snapshot: Sam and her father laughing about something long forgotten in Sam's shop, their reliable merriment binding them in friendship.

And another: Sam and Harrison assembled in the sitting room of Sam's house alongside the potter Maria Martinez, visiting from San Ildefonso Pueblo, New Mexico. Maria in her white boots. The conversation intense, signifying. "I sensed even then that here was an important Indian potter with two important craftsmen. My dad was excited about meeting Maria—that was a big deal to him. And Maria seemed honored to meet my dad and she knew Sam well." Then everybody shifting into the yard, sheltered from the caustic blaze of the summer's afternoon sun by a huge shade tree, watching attentively as Maria's grandson donned "this incredible turquoise headdress" and took the first jouncing half step and twirl into a traditional Pueblo dance.

How does the child of an important artist surrounded by other important artists catch and collect the fleeting impressions that one day will become crucial memories, even a basis for history? Children without notebooks and tape recorders: how much is lost? Thank goodness for the occasional child with a camera's eye.

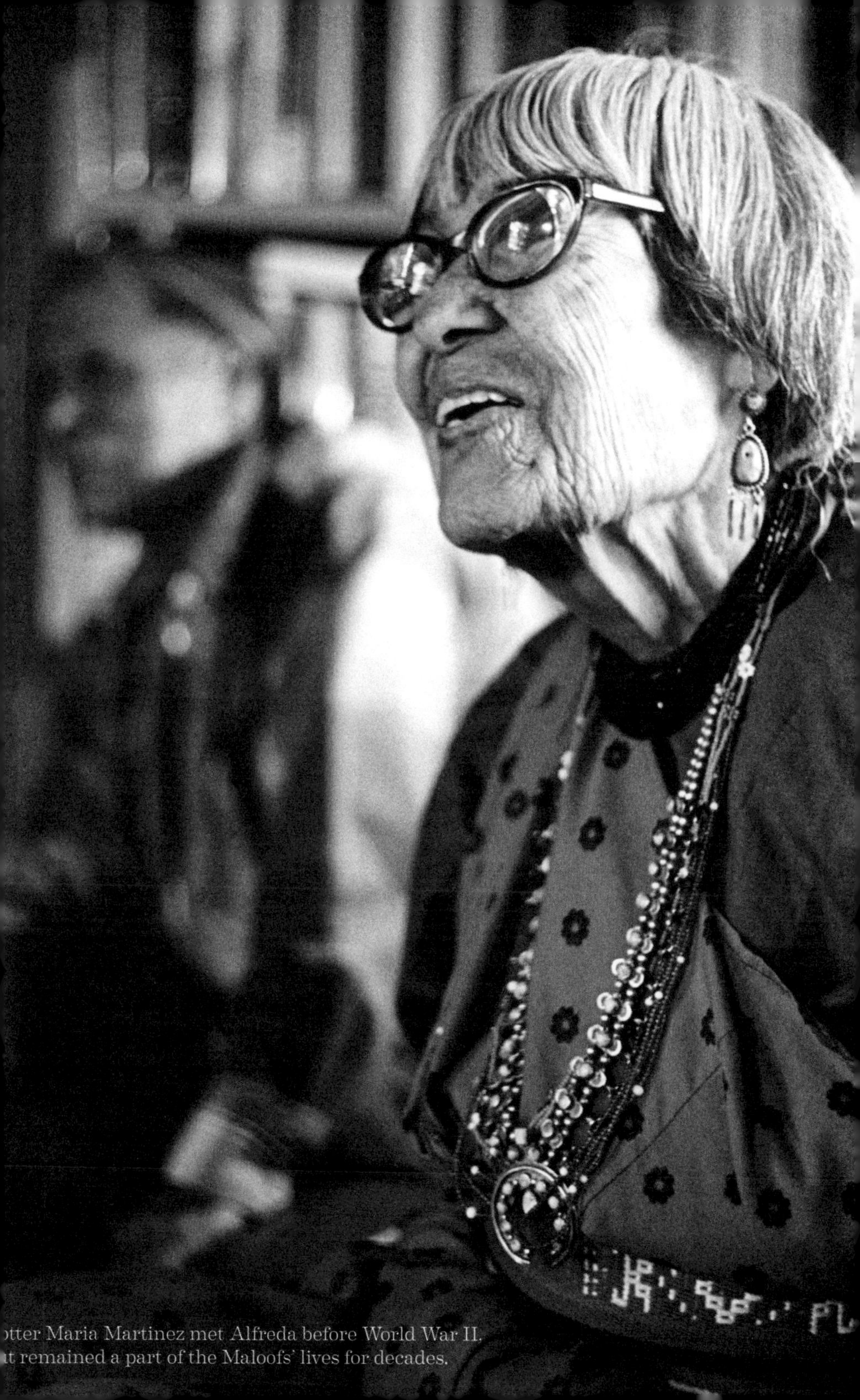

otter Maria Martinez met Alfreda before World War II.
it remained a part of the Maloofs' lives for decades.

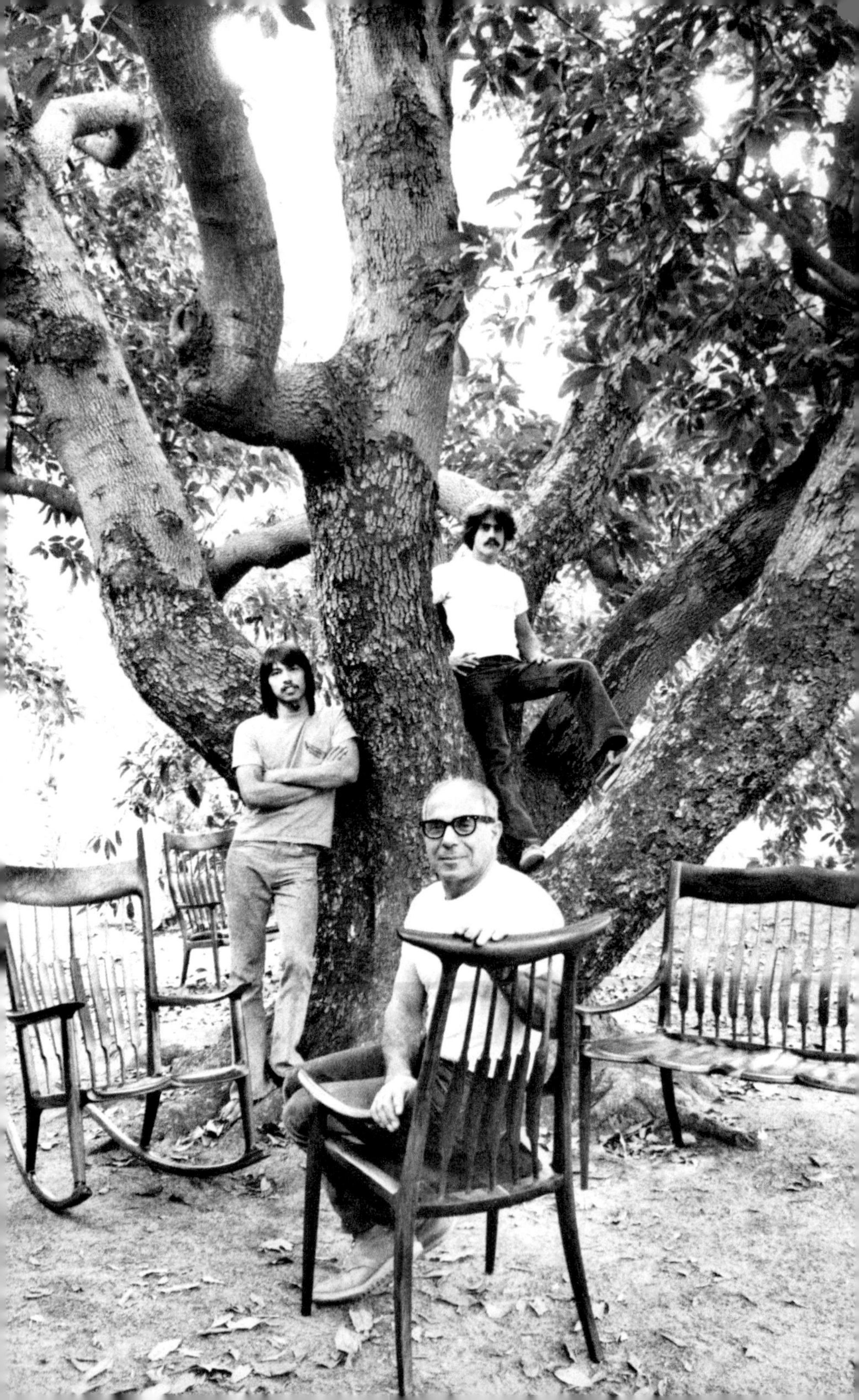

And her friends.

"We've got this Art Kids group we've made," explained Catherine. "The children of Claremont artists. There are thirty-eight of us. We have a party once a year. It's like a support group. We're dealing with similar things. The first year several people came from San Francisco and Arizona. Our last gathering, maybe eighteen actually showed up. Some stay in touch on our Facebook page. We've got a closed group so nobody else can see it. It's been useful to ask questions. How do you deal with such-and-such practical matter? But when we get together at parties, there are also conversations about what it was like growing up with artist parents."

The shape of their early lives had been privileged, informing, peculiar. Free and exhilarating to the point of risky for some young people whose reins ran too loose. And now, in middle age, the Art Kids find themselves freighted with a degree of unanticipated responsibility. How do the children of artists protect their parents' legacies? Guarantee that their names will still be mentioned, their work included in books and monographs? Ensure the care of identifiable pieces, sort through the requests from museums and scholars, correspond with patrons and enthusiasts, and help organize the odd exhibition that might still come around? The world in which they grew up has been dispersed by time, and it's left to the children of the artists to convey the form, texture, tone, and significance of an era that from almost every perspective must now appear golden.

s demand for Sam's
rniture grew, his nephew
asif Maloof, right, and
ul Vicente, left, helped
ep up with the orders.

16.

THE TRADES
KAY SEKIMACHI

We can envision Sam admiring the latest work by his friend, fiber artist Kay Sekimachi. "I want one," he says.

He might be referring to her narrow linen tubes suspended from the ceiling, some eighty inches long—part of the *Marugawa* series. Or her paper vessels. (They always delighted Sam; he owned three.) Or a small double-woven booklet with an accordion fold. You could stand it up or display it prone, depending on the available space.

"'I want one,'" remembered Kay. "That's all he'd say."

Sam expressed immediate, definite preferences. He trusted his instincts, seldom erring. "He had some of my best work," Kay acknowledged. Since 1972, when she married wood turner Bob Stocksdale, Sam's close friend for two decades, Kay had observed Sam's ability to step into a gallery or exhibition, scan the surroundings, and then quickly beat his way toward the most rewarding pieces. "You can train your eye," she said. "But you're also born with it. If you've got good taste, you've got good taste. My mother had good taste. It rubs off on you."

Among Sam, Bob, and Kay, the exchange of opinions and affinities over four decades proved continual, enriching. Each workday, Bob settled into the basement of the couple's home in the Berkeley flatlands, turning the sides of his birch, ebony, and ginko bowls to an improbable thinness, while Kay remained on the top floor, fixed to her card loom or manipulating fine fabrics and paper fibers by hand and tool. At the end of the day, they convened in the ground-floor living room to discuss their progress, reshaping each other's sensibilities with additive ruminations on the peculiarities of their respective materials and techniques. When Sam and Alfreda came to visit, often staying for several nights in the extra downstairs bedroom, the conversation inclined toward exhibitions, a little gossip, and furniture design and assembly, a new post of visual reference.

Sam had collected Bob's work since 1958, six years after they first met during a woodworking demonstration at the Los Angeles County Fair. Over the decades, he acquired thirty-four pieces, including salt and pepper shakers, a mortar and pestle, egg cups, and many bowls. Until his death in 2003, Bob made all of the ebony plugs Sam used in his furniture and often passed along wood supplies he deemed suitable for Sam's shop. In turn, Sam evangelized to the wider craft community about Bob's work, attracting private collectors and helping to secure his status as a fellow with the American Craft Council. When Sam won the MacArthur fellowship in 1985, Kay and Bob rejoiced not just for their friend, but also for the entire field of crafts. The attention and respect conferred on Sam radiated outward toward all creators of painstakingly beautiful handmade objects, the distinction between artisans and artists at last shrugged aside publicly. Kay, Bob, and Sam continued to communicate most deeply among themselves through the mediums of the things they made.

"Sam and Bob were very close," remembered Kay. "They didn't even have to talk."

So when Sam expressed his desire for one of Kay's pieces, she recognized the "need on his part to collect, and I just gave him what he wanted. No accounting whatsoever."

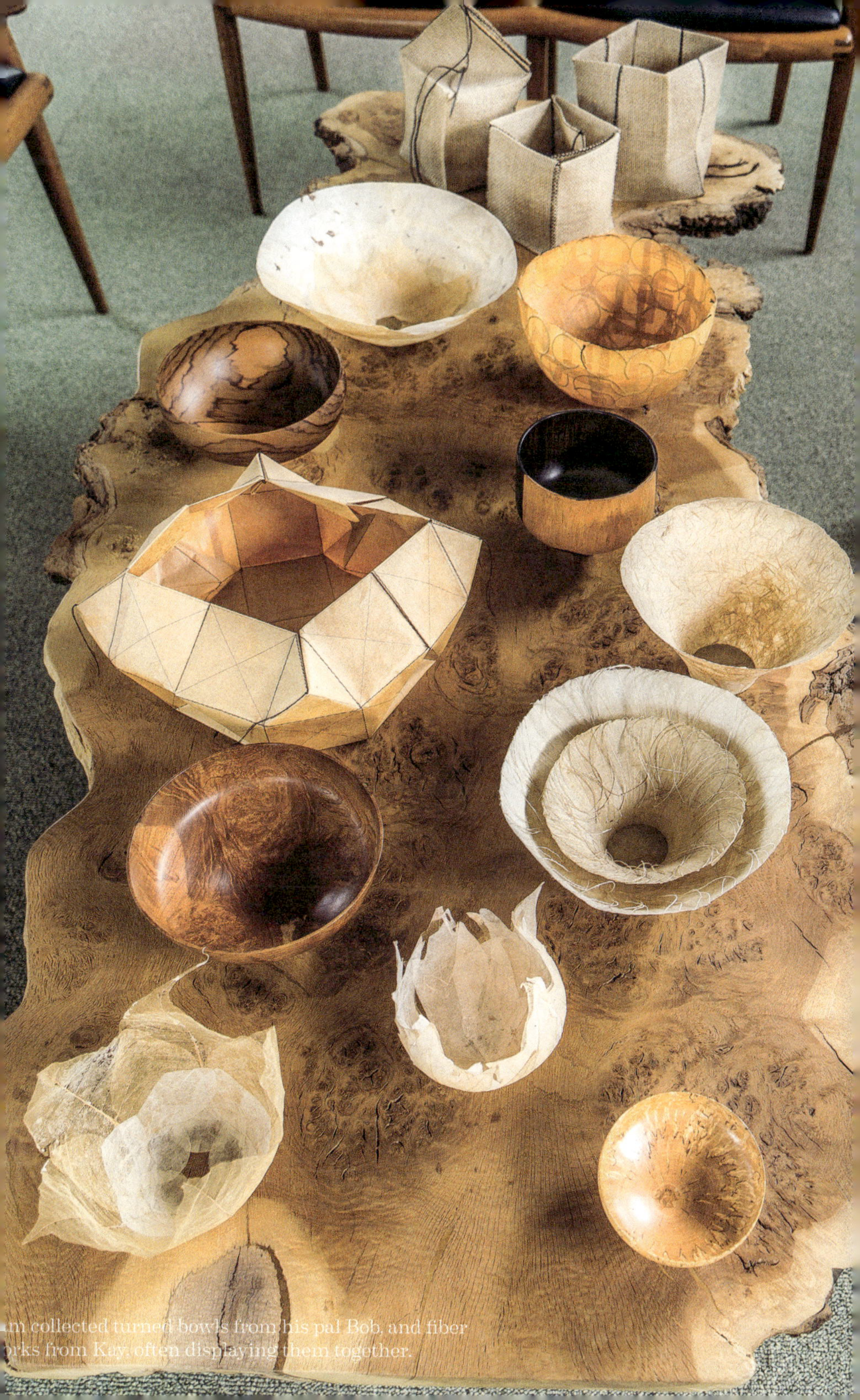

m collected turned bowls from his pal Bob, and fiber
orks from Kay, often displaying them together.

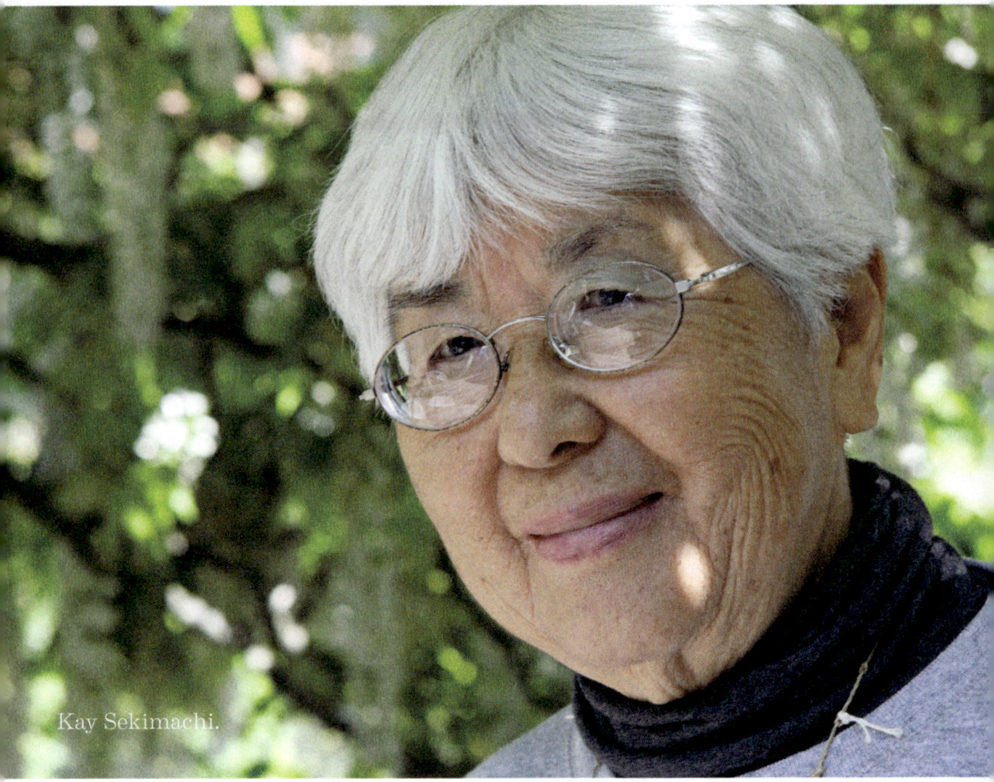

Kay Sekimachi.

No costs and value estimates tallied, no records kept. “We felt comfortable with what we got from each other.” Kay and Bob eventually completed trades for eleven Maloof pieces, including chairs, rockers, and tables—an unusual concentration of furniture for the 1,400-square-foot house where Kay continues to live and work. And with the sizeable collection, there came changes in daily use that could only be accommodated by a long, affectionate alliance.

“I’m using Bob’s walnut desk now,” said Kay. “Of course, Sam made it. But when I got my laptop, I needed to be able to plug it in.” Her stepson-in-law pressed his electric drill into the back of the desk and bored a rude hole to snake the cable toward a nearby electric socket. “I wonder if Sam would approve of this?” he asked.

Then there are the four chairs and small round table, four feet in diameter, that Sam gave to Kay and Bob as a wedding present. When the couple moved into their remodeled house, they wanted a larger tabletop that wouldn't be dwarfed in the natural light flooding into the front room's ample blocks of open space.

"If we wait for Sam to make the top," said Bob, "we'll never get it." So he made it himself and gave the original back to Sam. A table that's one of a kind, never to be replicated: the base by Sam Maloof with a top by Bob Stocksdale—the daring compromise a mark of their unfinicky friendship.

"And then," said Kay, "there is my chair. The one Sam made for me."

In the corner of her bedroom sits a straight-backed chair, conspicuously smaller than the standard design. Sam crafted the chair proportionately to fit Kay's short stature. "I used to think I was four-feet-eight," she said. "But no way now. And I'm still shrinking!" Kay stands in front of the chair and concentrates her gaze on the chair's legs. She has a confession to make.

"I don't sit in this chair that much."

She takes a seat.

Her toes dangle three inches off the floor.

"I never told Sam."

17.

SAM'S COLLECTION

Sam loved surrounding himself with beautiful objects. Pots, oil paintings, fiber art suspended from the ceiling and delicate weavings propped up on trays. Turned wood bowls, Navajo rugs and Oaxacan carvings, ceramic heads and stone faces, trunks and limbs culled from favorite trees, and bits of dried plants and twiggery he fetched from the grounds and gardens. Enameled abstractions on copper plates, sketches, prints, and watercolor landscapes, baskets and beaded moccasins, jewelry, porcelain, stoneware vases, sculptures large and small. He could mix folk art sold off the road in Kentucky with museum pieces; he arrayed the ceramics and paintings of his famous peers and predecessors—Gertrud and Otto Natzler, Emil Kosa, Milford Zornes, and Betty Davenport Ford—alongside student work that intrigued him. What mattered most—and it mattered exclusively at the start of his collecting in the fifties and sixties—was delight, curiosity, provocation, interest. Unlike some famous artists who disregard their contemporaries' efforts and keep their homes free of others' work, Sam found stimulation and nourishment in art that bore not the slightest resemblance to his own.

One of his friends, who remembered the collection in its first vestiges, likened Sam's acquisitive instincts to "very controlled hoarding." Everything he admired, he wanted. Oftentimes, he amassed multiple iterations; better ten Persian rugs than one. If he appreciated a particular artist, he could purchase or trade his way to sizeable stocks—eight paintings by Sue Hertel, twenty-one ceramics by Harrison McIntosh, thirty-four of Bob Stocksdale's turned-wood pieces.

Sam "needed" new art on a constant basis. Not that he pursued a strategy. Sam selected each piece for the pleasure it brought him, sizing up its agreeably unstable standing amid the expanding assembly. The collection radiated none of the self-conscious delicacy and self-regard associated with the snootier forms of connoisseurship. Sam's choices proved too hardy and idiosyncratic. He purchased and traded to sustain his collection as a work in progress; he never aimed for an end.

Sam could seldom leave any arrangement intact; he tinkered relentlessly with the pieces. The introduction of a new vase might call for the reorganization of a room. Nobody but Sam had any idea about the disposition of the entire collection at any particular moment, and he alone could instantly place his hands on a desired object. The norm was purposeful, animate flux; everything, ideally, communicating with everything else.

By the mid-1980s, Sam's collection had earned a reputation for its generous scope of Southern Californian postwar arts and crafts. It certainly ranked as the best private grouping of the Pomona Valley artists. Younger painters, ceramicists, and sculptors vied for inclusion; membership in the Maloof collection could boost a career. For years, Sam could be routinely seen sweeping through local galleries and exhibitions (he was partial to Cal State San Bernardino) that featured the work of ambitious students. While he still loved most of what he purchased, he also pursued a secondary goal: he wanted to encourage young artists he thought promising,

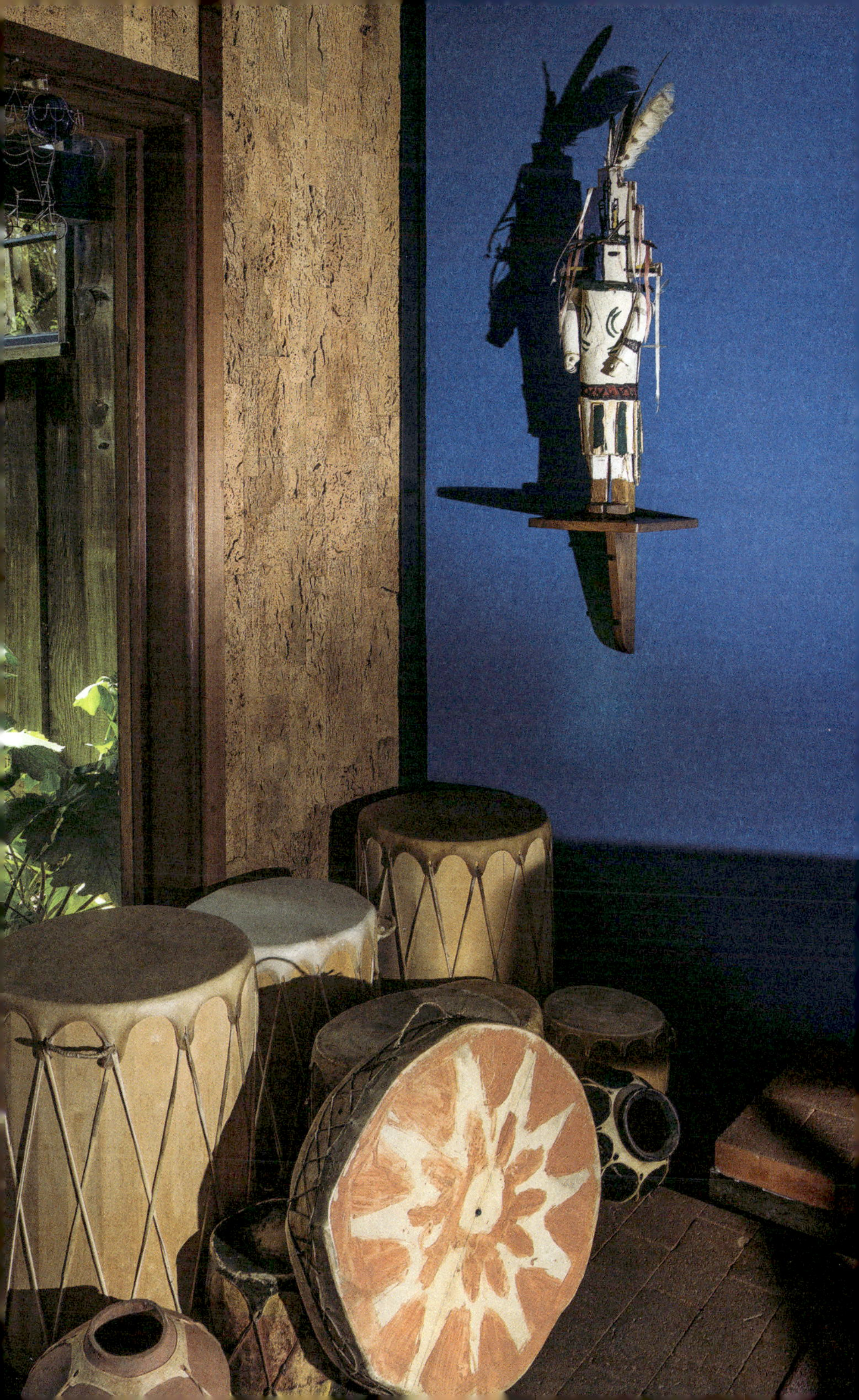

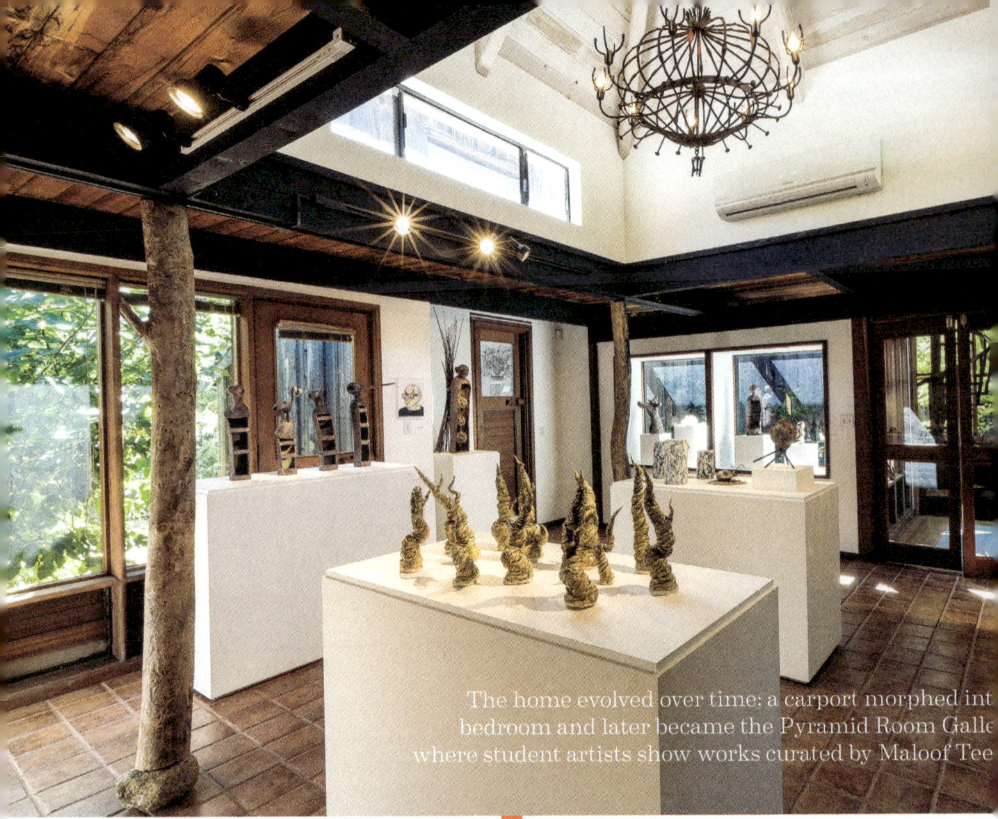

The home evolved over time: a carport morphed int
bedroom and later became the Pyramid Room Galle
where student artists show works curated by Maloof Tee

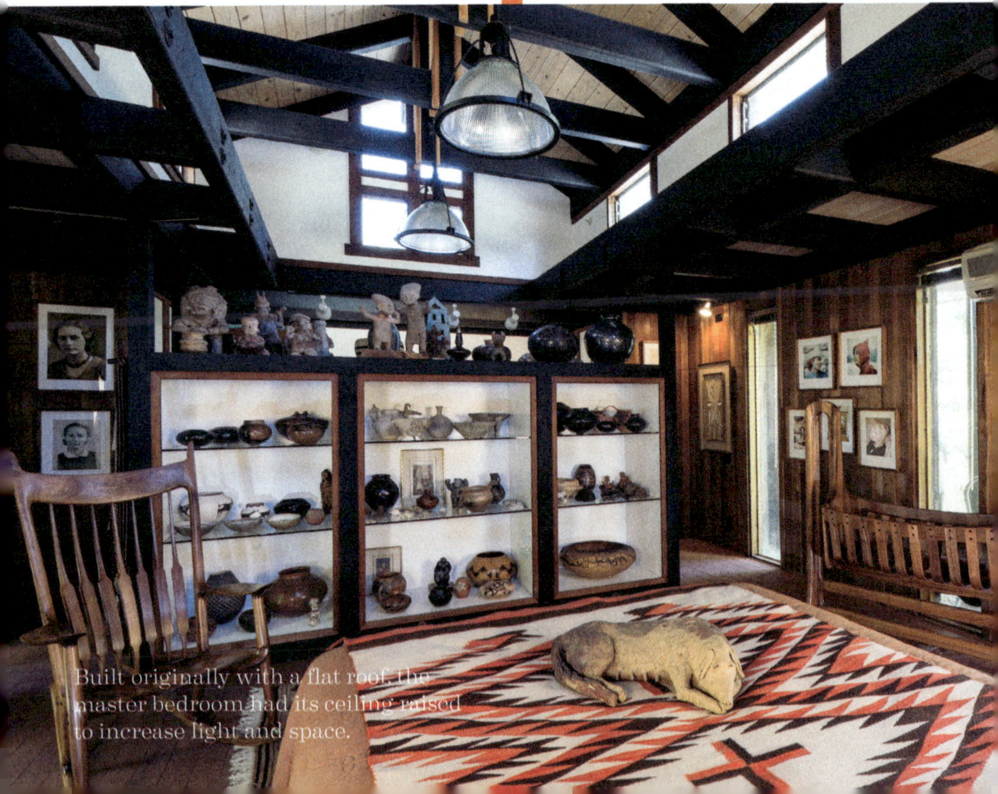

Built originally with a flat roof, the
master bedroom had its ceiling raised
to increase light and space.

hardworking, itching to stretch boundaries, and probably barely getting by. His enthusiasm and catholicity of tastes made him a patron in the best sense of the word.

At the turn of the twenty-first century, with his gallery roaming unabated and his acquisitive drive still keen, Sam occasionally became the recipient of unsolicited gifts. Artists he had never met or even heard of forwarded their work with a friendly suggestion that it belonged in his collection. Sam accepted these pieces, though they did not necessarily secure a coveted place on prominent shelves or walls.

With Sam's death in 2009, his personal, often impulsively acquired collection began its gradual shuffle into institutional order. The formal presentation of some two thousand pieces valued at between $5 and $6 million required selectivity. Sam had kept his house "chockablock full of work," as one friend recalled, "absolutely filled." With the construction of the education and archival storage centers, the foundation could now rotate work in and out of public view, highlighting themes and eras. A reordering of the upstairs gallery, for example, prompted the removal of at least thirty pieces, including three paintings. (Every shelf had previously contained seven to nine works with many others spread across the floor.) For a home, it could seem cluttered. In a historic residence hosting a constant trail of visitors, it shouted chaos.

In 2011, Linda Apodaca joined the foundation as the collection manager. Her first task: determine exactly what the collection contained. An earlier effort at constructing a database had identified three hundred objects, though the method employed an eccentric numbering system out of step with standard museum practices. Linda completed and cleaned up the inventory. She appended photographs, appraisal prices, and background information to the records, along with notes specifying the location of each piece.

Dependable records cleared the way for other responsibilities. Cleaning. (The objects' immense amount of surface area attracted a commensurate quantity of dust.) Security. (Today only the manager moves pieces around, and that's done with an eye toward safety and recapturing the variety once assured by Sam's persistent rearranging.) Order has also facilitated outreach. Given its abundance of fine work, the foundation now eagerly loans out pieces for exhibitions nationwide. To date, at least twenty museums and colleges have bolstered shows and displays by borrowing from the collection, with more arrangements pending.

Perhaps the most complicated task has been incorporating the spirit of Sam's collection into the house's rekindled identity as a public museum. Sam rejected excessive systematizing; the collection was "Just my stuff...." But unlike many private collections, Sam's numerous pieces could remain together precisely because they had a home. Indeed, the house remains an inseparable part of the collection, the artwork arranged in counterpoint to a structure that is itself a work of art. Sam—of course—is the missing ingredient. Keeping his sensibility alive and true remains the challenge.

To present a strong collection that reflects both Sam and Alfreda requires a degree of artificiality, a rejiggering of the past. Archival photos show the house with far less artwork on display in its earlier years; the further you go back, the fewer the pieces. The curatorial response has been to pare and shape the collection to look more like it did thirty years ago. Alfreda's influence in terms of Southwestern Indian art can be viewed in virtually every room. There's the huge rug collection, the mass of drums on the floor, a display of black-on-black pots by Maria Martinez with photos of her family posted on the wall. (And another pot in the same style—and to the untrained eye indistinguishable—made under Maria's tutelage by Alfreda.) Alfreda's cookbooks fill the kitchen shelves, and family photos punctuate occasional blank space along the walls. From another perspective, the house and collection seems a diorama of everyday domesticity, even as it evolved into an inadvertent, informal salon of midcentury modernism.

The sheer quantity of art by Sam's friends underscores the dailiness of the household collection. Three pencil drawings by Jim Hueter posted over the sitting room settee. Two sculptures of *Men with Hats* by Larry White fit into the nook of the tree house room. Near the alcove at the top of the stairs, Phil Dike's painting *Spear Fishers #1.* In the mezzanine gallery, atop Sam's Tasmanian eucalyptus burl table, ceramics by Harrison McIntosh and Rupert Deese. In the showroom, set upon Sam's natural-edge walnut burl table, wooden bowls by Bob Stocksdale and their fiber counterparts by Kay Sekimachi. Sometimes the origin of an acquisition is expressly known. The painting hanging to the left of the fireplace, *Convoy to India,* was a wedding present for Sam and Alfreda from Millard Sheets. But the provenances of many pieces remain mysterious and probably impossible to unearth today. While Sam engaged in numerous trades with other artists throughout the sixties and seventies, the details often passed unrecorded. So many people who entered the house, sharing meals and conversation, building their lives and careers in tandem, spent their days and nights creating sculpture, paintings, ceramics. Trading one beautiful object for another simply made sense. And at that point, nobody thought about Sam's collection in terms of posterity.

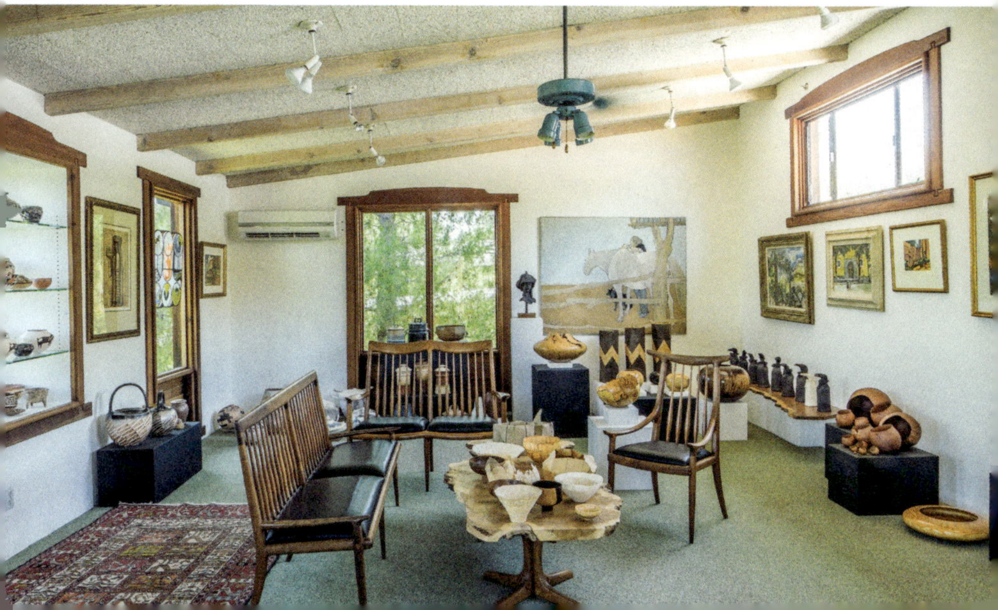

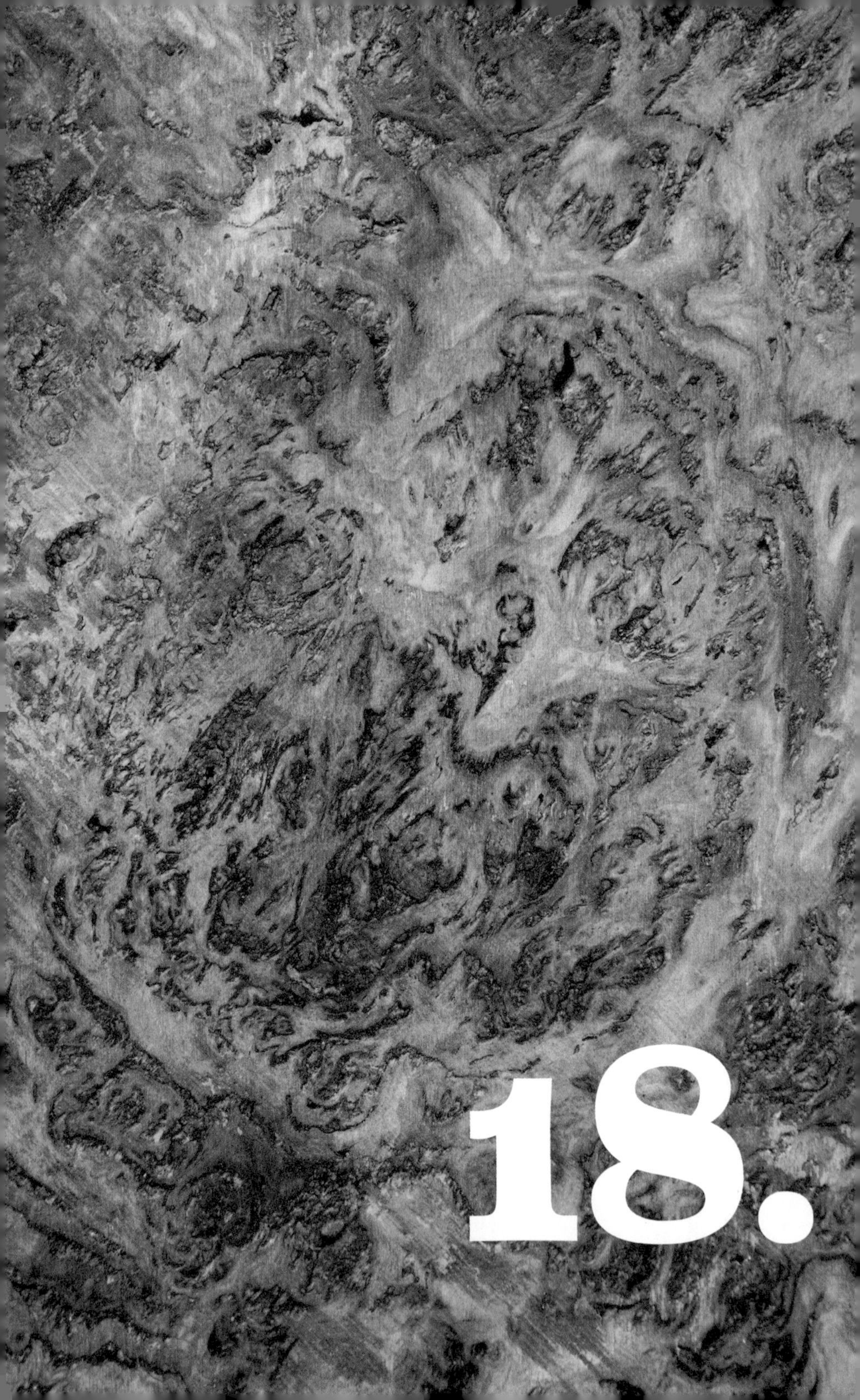

18.

BEAUTY
JOHN AND DEBBIE STARR

In 1969, Sam Maloof began a series of twenty-eight straight-backed rocking chairs as a commission for the Los Angeles chapter of YPO, the Young Presidents Organization—an international network of corporate executives whose 22,000 members ascended to top leadership before the age of forty.

"The tradition," explained John Starr, chairman of UltraSource, a Kansas City producer of equipment and consumables for the meat industry—and YPO member —"is that when you turn forty-nine, they kick you out. It's called 'forty-nining out.' And they give you a rocking chair that says YPO on it."

The Maloof rocking chair as gift and provocation. An acknowledgment of early achievement—and a gently mocking reminder that the recipient must not quietly tip and totter his way through the forthcoming years.

The YPO commission furthered the reputation of Sam's signature rockers. For John Starr, it's a telling, touching coincidence.

"It connects two parts of my life," he explained. "I'm very active in YPO and I love collecting Maloof furniture."

This morning, John and his wife, Debbie, have arrived at the Maloof workshop to select wood for a new commission: a table and sideboard, design to be determined.

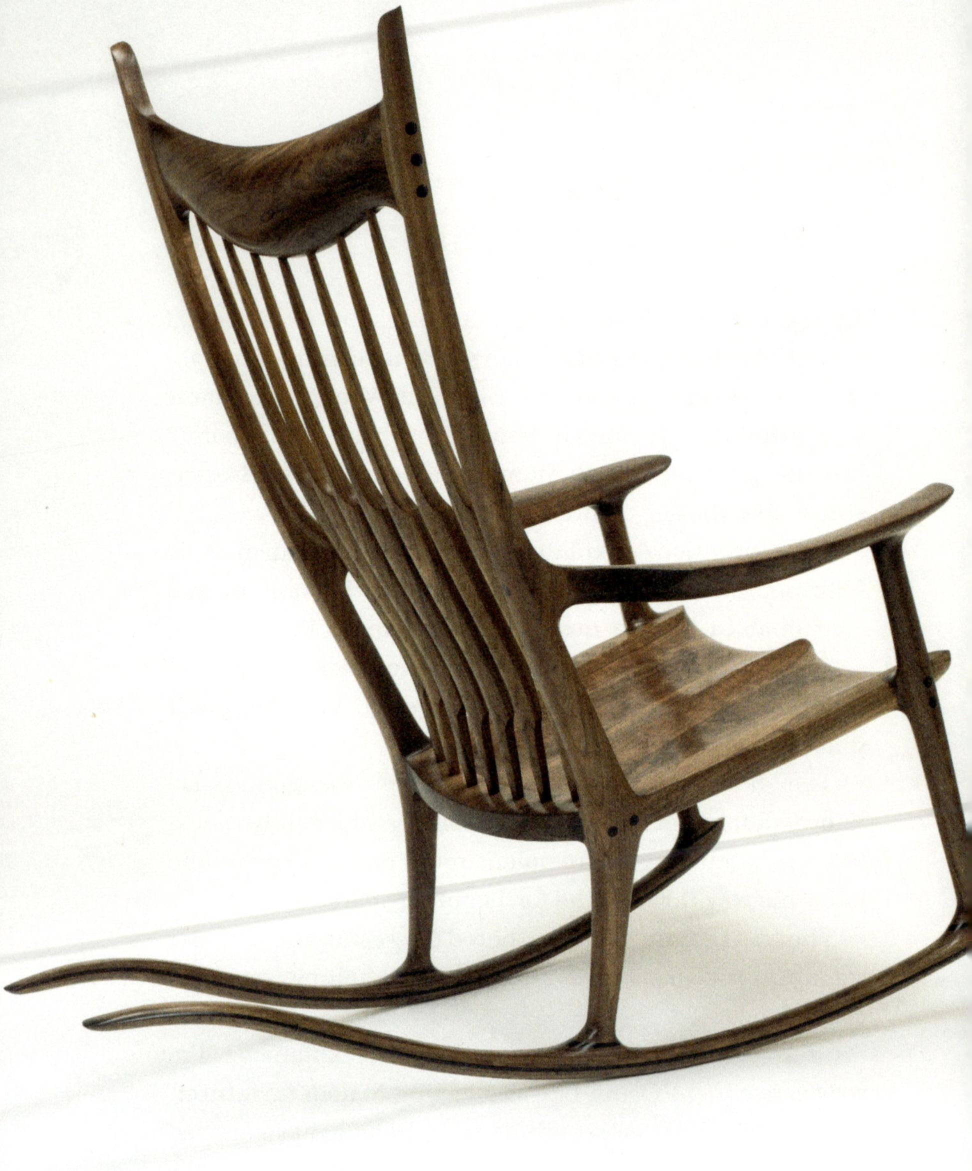

As the Maloof rocker evolved, Sam's signature piece became more sculpted than its predecessors. A hardwood seat replaced the earlier upholstered versions.

"It's a big mistake to try to do this on the phone," he said. "I have to see the wood. They've got so much here. They could pick out a piece that would be beautiful. But there might be something that I'd like a little bit better."

Years of collecting Maloof furniture—the Starrs own at least eighteen pieces—have taught John that he responds intuitively, viscerally. What he sees and experiences crystallizes desire: the wood and the wanting articulate possibility. "It's about my style. What shade, what color, what kind of edge."

It's another coincidence that Debbie's parents live less than a mile down the road from the Maloof compound. "I grew up here and I didn't know about Maloof," she said with wonder. "I used to walk by all the time and not know what it was."

As John suggests, "People who know wood, know Maloof." But we're talking about a small slice of humanity, people who have invested some part of their lives in learning to appreciate useful objects artfully fashioned from wood. Fewer still are those visiting the Starrs' home who can distinguish their Maloof tables from their Nakashima tables. Or sort their George Nakashima pieces from the work of his daughter, Mira. Or identify the ziricote, the Tasmanian eucalyptus, the Indian rosewood, and the odd cuts of maple with their flat edges pressed up against the wall—woodshed remainders that John stumbled over and immediately purchased in bulk for their beauty and imperfections.

Yet the Maloof pieces communicate forcefully, whether or not visitors care about furniture.

"People walk into our home," said Debbie, "and it could be the postman or a friend who's a billionaire. The first thing they notice are the Maloof chairs. It has been shocking to me that people who you'd never expect to recognize how amazing these chairs are—well, it's the first thing they see. There's a lyricism to the way they're constructed. They make you want to sit down. And when

I sit in the chairs, my favorite thing to do is put my hands behind them and feel how it's all connected."

"Yeah," said John, "tactile. When we got our dining room chairs from Sam, which took—what, eight years?"

"Because you negotiated such a low price for them." Debbie laughs.

"That's not the reason at all. I did. But that's not the reason. People just had to wait. Years. When they finally did arrive, I sat in one, and it just...grabbed my butt. I called Sam immediately. I said, 'How did you design that chair so it held me in perfectly? How'd you know my posterior?' He said, 'I designed it for *my* posterior.' It turns out that his posterior fits most people's posteriors."

The lyricism of the perfectly posteriored chair.

"To me, fine art—whether it's music or visual—you don't get tired of it. You experience its beauty over and over and over again." Flying west from Kansas City earlier in the week, John clapped on his ear phones to submerge himself in Beethoven's Piano Sonata no. 23, the Appassionata, one of the most challenging and complex of Beethoven's piano works. John calculates that he must have listened to the sonata at least five hundred times. "I still love hearing it. I get calm. It's a stress dissolver. When I sit in a Maloof chair, it's a stress dissolver. My focus turns to the experience of sitting, to the curves. To wondering, How did he think of that angle? And I'll start feeling the wood, the edges."

"I don't think of Sam's work as a visual art," said Debbie. "I think of it as the way music surrounds you. Encompasses you. With a painting, you see it, and it's separate from you. But because you use Sam's art, it's not separate. We rely so much on our visual sense that we sometimes forget that we know about the world in some ways more intimately through other senses."

And yet it's sight that first draws our attention, startles us.

John remembers the first piece he commissioned from Sam. "He walked me through his museum, if you will—his house. And I saw a Tasmanian eucalyptus table. Little holes, thousands of these little holes. 'Do you have any more wood?' He had one more piece, a really rare piece. 'Put my name on it,' I told him. 'I don't care when I get it.'"

"It's still one of my favorite things to look at and feel," said Debbie. "It's a table, but because of the shape of the burl, it's as though you're looking down at a tree. Full of crevices, and the base like a trunk. In the middle of our living room is this amazing table, round and about four and a half feet in diameter, that looks like a tree."

And what does the presence of this table, the tree in the living room, invite admirers to do?

"Protect it," said John. "If people put a glass of water on it, it leaves a stain and—"

"It makes me want to drag my fingers across it," said Debbie. "It looks like it's being invited to be touched."

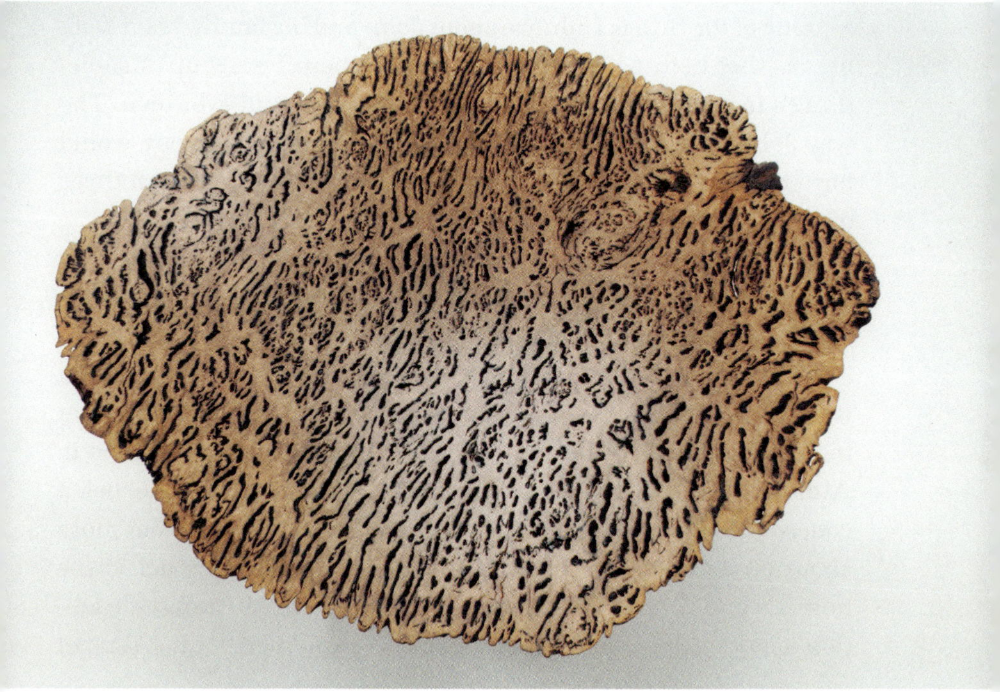

Art as an aspect of daily living, routine, and use. John grew up surrounded by art acquired by his grandparents, collectors at the tipping point of early modernism. His parents expanded the collection while his mother, Irma Starr—an esteemed ceramicist whose own work has been acquired by the Nelson-Atkins Museum of Art and the Smithsonian—collected rare Japanese ceramics. John recalls using them daily while he was growing up.

"I can't remember anything breaking, but if it did it could have easily been a hundred grand. For a teacup. The million-dollar plates we only used occasionally—when my sister came home. The teacups were for every day. We used to eat off Lalique glass plates. Sam's approach was the same. 'There's nothing you can do to these tables that can really hurt them. You can always sand them down and re-oil them, and they're back good as new.' Well, you can do it, I thought, because you've got a shop. I'm out here in Kansas City. That's why I don't have a dining room table. I like things pristine."

"One of the things I admire about John and his family," said Debbie, "is that nothing is beyond bounds. Where I grew up, if something's in a museum, it's not something you would ever own. The way John grew up, if something's in a museum, somebody would have had to have made it—and we better go meet them. All three of our kids are like that."

Their son, for instance, attended an exhibition in Kansas City about the St. Louis World's Fair. A piece of furniture caught his eye.

"Boxes on boxes," said John, "recesses and doors. He thought it was the coolest thing and he decided he wanted a wall like that in his bedroom at home. So he designed it, had somebody build it. Almost $70,000 to build—I had no idea at the time....But he took a concept he saw in a museum, talked to the curator, found out more about it, worked with a furniture maker, created a model to see how it worked, tweaked it, and then executed it. My family's tradition is not to go to auctions or galleries," explained John. "You get a name"—an artist, a curator—"and you visit them. You have the confidence. If you have the interest and you can afford it, it's okay to do that."

"Life's an unfolding of micro-interactions," agreed Debbie. "You can have a big idea, but it's the micro-interactions that make it happen."

"One of the things I truly enjoyed about buying Maloof pieces," said John, "was my time with Sam."

"He was like a little kid when he was going to see Sam."

"We'd stay at Debbie's parents' house. I'd come up here and look at wood, see what they're doing. Talk for hours with Sam. I loved my time with Sam. Loved that he loved In-N-Out Burger because it's my favorite, too."

"John, I think it rooted you," Debbie said. "It gave you back something you missed. In some ways, your connection with Sam allowed you to reconnect to a really important part of your life that you wanted to be ours."

Their collection, their collaboration. Growing into a mutual sympathy for the chairs and tables, the credenza and headboards, acquired through daily use without sacrificing the prerogative of astonishment.

"I just visited a very major private collection," said John, "with a full-time curator just to take care of this guy's art. It's like a museum. Only 10 percent could be shown in this big house. Constant rotation." He turned to Debbie and wagged his head. "I don't want to have to start a museum just to hold my art. I've got to live with it."

19.

THE DEAL
LINDELL MARSH

Sam spread a map across the lunch table as Alfreda cleared the dishes. His fingers traced the course of the proposed freeway extension. Twenty-eight miles of eight-lane expressway from La Verne to San Bernardino. Its route ran straight through Sam's property.

"But it's probably just talk," Sam said more than once.

Sam couldn't believe it would happen. Refused to believe. State planners had been discussing the freeway extension for years. Even in the antic rush of Southern California, public works projects could plug along sluggishly. Sam had long regarded the prospect of the freeway as more an annoyance than a threat. In truth, he couldn't accept that something so terrible as the destruction of his house and workshop, the paving over of his grounds and garden, the uprooting of his family and the disruption of his career, could be callously planned and executed by the State of California. The shape of his life, as he preferred to view it, did not allow for disaster of this magnitude.

Lindell Marsh, his attorney, could not agree. He didn't know if the freeway would prove inevitable, but it seemed likely. He scanned the map's layout of freeways heading north. Traffic engineers had spaced them at roughly equal distances. The San Bernardino linkage fit logically into the framework. Many commuters would have deemed it long overdue. Sam needed help.

As an attorney, Lindell specialized in land and environmental law, a growth field in a booming state riven by competing visions and interests. He and Sam had first met in 1969, when Lindell commissioned a rocking chair to commemorate the birth of his first child. Later, while working with an architectural firm intent on revitalizing Hollywood, he had admired a partner's desk equipped with two sides and a table in the middle—ideal for unfurling maps and plans and bearing down in concert on the particulars. He conferred with Sam, and together they designed a similar piece for Lindell's office. A set of chairs and a round conference table followed. Each piece required hours spent in Sam's shop—the conversation stimulating, with hearty meals provided by Alfreda. Eventually, Lindell felt that he could call Sam his friend.

He considered the options now facing them.

Their strategy hinged upon a single, untested proposition. Federal law compelled state and county governments to enter into a mitigation process when highway projects threatened historically significant properties. Sam's house had been deemed eligible for listing on the National Register of Historic Places, automatically conferring significance on the residence. But just how the mitigation process would proceed seemed far from clear.

"At the start," said Lindell, "I strongly believed that you could not replicate Sam's place. It was a matter of figuring out how to keep it."

The layers of authority and the number of agencies involved in the deliberations almost guaranteed a muddle. At the top level, San Bernardino Associated Governments (SANBAG), an amalgamation of staff and elected officials from the county's twenty-four cities, assumed overall responsibility for the freeway extension. The California Department of Transportation (Caltrans) played a key role in formulating state plans and allocating funds. Finally, the Federal Highway Administration asserted its own demands on the state, which ultimately played out at regional and local levels. When Sam, Alfreda, their son, Slimen, the City of Rancho Cucamonga, a team of construction experts, and other government agencies eventually entered into the conversation with Lindell and his associates, the meetings could involve as many as forty individuals.

Fortunately, Lindell had been preparing over the course of his career for precisely this kind of contentious, unwieldy assembly. For years, he had worked amid a whirl of local, state, and federal agencies in a variety of witheringly complex negotiations involving California's disputed coastal regions, wetlands, and designated open spaces. What others might reasonably perceive as a splay of impenetrable barriers, a bureaucratic hedge against satisfactory ends, he regarded as an opportunity to fine-tune the public good.

"I was aware of the significant challenges of getting agencies to work together," he acknowledged. "But my deep interest is collaboration in governance. I don't litigate at all." After conducting the necessary research, his legal team set the tone for coming deliberations with a bold declaration: they had no interest in attempting to stop the freeway. "But Sam's house is eligible for the National Register of Historic Places," Lindell cautioned the county, state, and federal interlocutors. "You need to deal with that."

When SANBAG asked what Sam and his backers wanted, Lindell had to admit that he didn't yet know. Models for handling such a sprawling, unique property did not exist. They would need to invent the model.

"Gary Moon, from SANBAG, was critical," said Lindell. "He perceived that we were sincere." Moon invited Parsons Brinckerhoff, their agency's engineering firm, to some exploratory meetings. Lindell reacquainted himself with the deputy to the state secretary of transportation, who reached out to the state historic preservation officer, who in turn dispatched an architect to con-

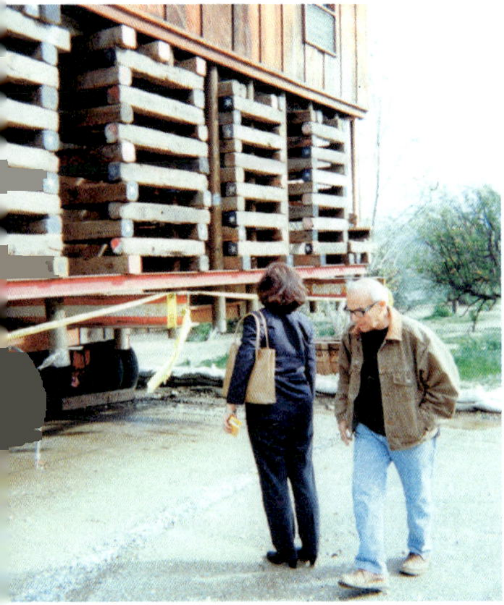
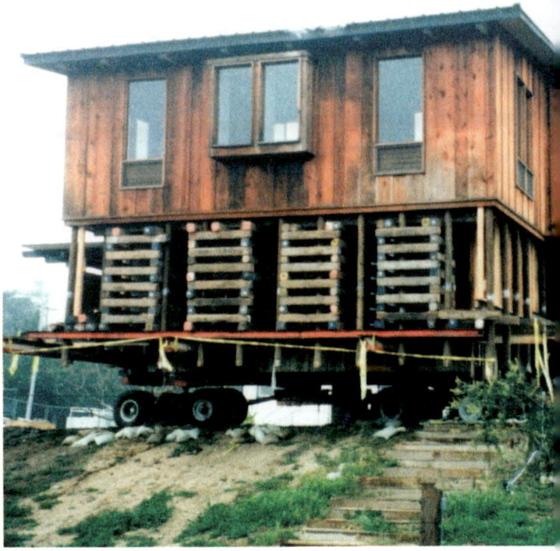

Sam's house prepares for the move.

firm the property's qualifying status. Meetings, dead-ends, more meetings, new possibilities—somehow they kept going. The key to their perseverance: avoiding any mention of "negotiations," a term redolent of winners and losers entangled in a conclusion that nobody actually desired.

"We were really saying, 'You help me, I'll help you,'" remembered Lindell. "People started putting themselves in everybody else's shoes. We built a level of trust."

Over time, the group managed to disrupt the old order, undermining the habits of competing interests and adversarial isolation. Learning to express their goals and limitations honestly, they grew into mutual respect, even sympathy, viewing one another not as selfish obstructionists or rigid bureaucrats, but rather as people of good will in search of a common strategy.

"Had we gotten into a fight," said Lindell, "Sam would have argued how important his house was and they would have argued for the importance of the freeway." Almost certainly, Sam would have lost the fight, and everybody would have lost the house.

"What we did was look at alternatives. 'Let me show you what my thoughts are to solve this problem, and you show me yours.' That way, we worked together to improve each other's ideas."

Some early proposals fell flat.

One alternative involved swerving the freeway, taking one hundred and twenty houses in its wake. Sam refused. "I can't do that," he said. "I can't be responsible for throwing those people out of their homes." Another plan championed the installation of a sound wall, shielding Sam's home and shop from the bustle and clamor of the freeway. The state would absorb half of Sam's property, reducing his six acres to three, and insert the barrier along a calculated acoustical rim. But rather than liberating Sam from noise and disruption, the sound wall evoked visions of penitence within an elaborate cell.

Finally, Gary Moon told Lindell that he wanted to explore the possibility of relocating Sam's house, shop, and other structures—the first time this radical notion had been broached. "Would you be willing to participate?" he asked Lindell.

"I suddenly wondered, 'Should I do this as an attorney representing my client?' I didn't want to be hamstrung. But there was trust. I thought he was making the suggestion in good faith." Lindell's own conception of what might be possible had altered considerably over time. "It was a good lesson for me that when you get into these things, you don't know enough to know what the best solution is. It becomes a creative process." To find a solution that could satisfy everyone, the group "needed to transcend order. That's what you're trying to encourage—a willingness to play so you're not bound by your initial assumptions."

Knox Mellon, who had formerly served as the state historic preservation officer, joined the group, providing a large measure of technical expertise in preparation for the move. Sam and his family attended sessions now swollen with agency staff. "My way of dealing with these processes," said Lindell, "is whoever wants to be

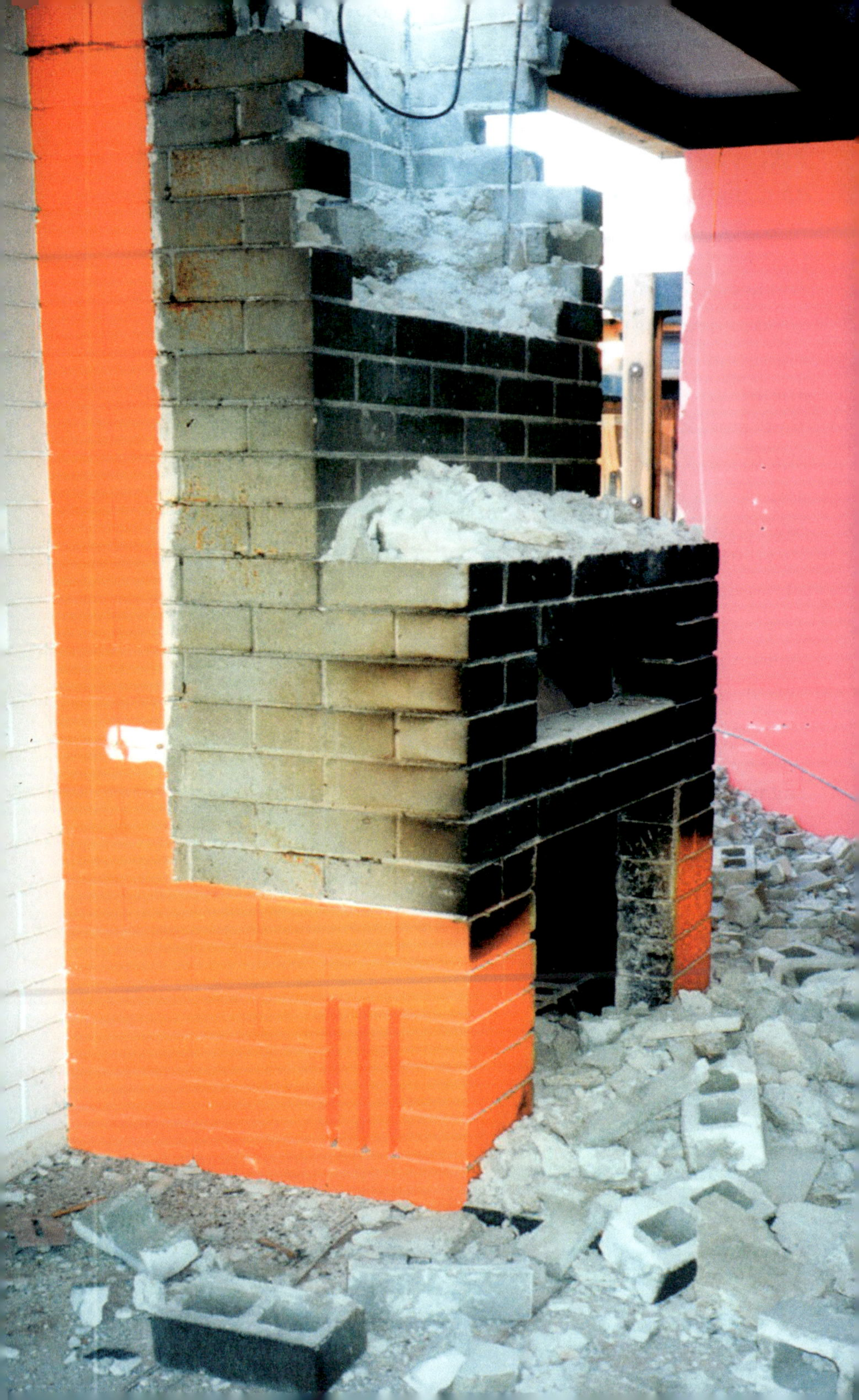

there can be there. There wasn't secrecy. It was open." Eventually, the group hammered together "this conglomeration of elements. A place for Sam to work. An interim house—which grew in my mind from two thousand to four thousand feet, and which Sam designed. Then the idea of the foundation that would hold the land and provide part of an endowment. The structures would be dismantled into eight pieces, moved at night, and reassembled two and a half miles to the northeast in Alta Loma. Sam's home and workshop became part of a nonprofit cultural center and museum.

If that plan had been put on the table up front," Lindell emphasized, "it never would have happened. People had to come to it. Including Sam."

Perhaps Sam most of all. Battered, depressed, combative, and only reluctantly compliant, Sam proved "very difficult to deal with." He inserted himself into the old property as it was being dismantled and then slipped onto the new construction site, his aggravating demands and potentially dangerous hand-waving interventions aimed at halting the destruction of his beloved trees and overseeing the deployment of heavy equipment. Eventually, SANBAG banned him from the property, a humiliating sideline to the supposedly triumphant conclusion.

Sam's relocation had proved a success. Preservationists would regard it as a prototype for future endeavors. Planners could point with satisfaction to the results of their skillful good will. The new site would now stand in perpetuity as an unequivocal declaration of Sam's importance as an artist, his house an ingeniously wrought testimony to his scope.

Yet for Sam, the deliverance of his irreplaceable home would always contain a measure of catastrophe, a reminder of the misfortune that had made the nearly impossible feat necessary. In his eighties, recently widowed and bereft as a man might be, they had ripped up the evidence of his life by the roots, shipped and reassembled the pieces a worldly two and a half miles away, and now everybody expected him to somehow grow into whatever came next.

20.

HISTORY
KNOX AND CARLOTTA MELLON

"Every time I walk through this room," said Knox Mellon, "I look at those pieces, and I think of Sam. And I walk into this room dozens of times each day."

He directed his gaze at the two chairs flanking the end table in his living room. A small-back chair and a somewhat shorter rocker.

"Sitting in here in the bright sunlight," Knox continued, "these chairs come alive. It's like the spirit of Sam coming back."

When you're perched on a hillside overlooking a gorgeous jag of coastline in Carmel, California, the emergence of the sun from the morning fog qualifies as an event—as befits a visitation from Sam. And it's an ideal location: the home of his longtime friends, Knox and Carlotta Mellon, amid their collection of African art and the towering wood totem figures from New Guinea, among their menagerie of rescue dogs and cats equipped with the names of Greek, Roman, and Scandinavian gods (and where Thor, the dachshund, growlingly worries his bone at the foot of the rocker).

"More than remembering him," suggested Carlotta, "I think of how he approached craft. There was no ostentation whatsoever. People would visit his house and they'd ask him, 'Can we sit on the chairs?' 'Of course,' he'd say. Sam allowed his dogs to sit in them."

It's a reverie of time-tempered friendship, Knox and Carlotta cognizant of how chance brought them all together. Appreciative to have altered and enlarged one another's lives.

Knox had just been mustered out of the army, part of the long road back in the Pacific War. Two days after his seventeenth birthday in December 1942, he impetuously left high school to join the marines, who turned him down because he turned out to be color-blind. The navy, too, rejected his application. "The army," he recalled, "didn't care." Knox traded his senior year in high school for three years in the South Pacific.

Back home in California, he quickly earned his diploma and followed a girlfriend to Scripps. One afternoon, he wandered over to Pomona College. The sign read: Admissions, Second Floor. Knox climbed the stairs, tapped at the office door, and introduced himself to the director. "We chatted away, got along splendidly. 'Why don't you come to Pomona?' he wondered. I asked what I had to do to get in. 'Write me a letter,' he suggested. Which I did the next morning. Three days later, I received an official announcement approving my admission."

Having just lived *The Iliad* in the jungles of the Malay Archipelago, Knox opted to study ancient Greece in tranquil Claremont. Back then, the small town was bordered by serried rows of bountiful orange trees, and when somebody dragged a paper scroll across the screen of the only movie theater in town, proclaiming "Frost—32 degrees," the room half emptied because so many in the audience pegged their livelihood to the orchards.

Knox's girlfriend worked with a sculptor at Scripps, and Knox tagged along to art events, mixing with the young students and instructors filing into the department shortly after the war. There, he met Sam—a casual acquaintance at first. Knox also got to

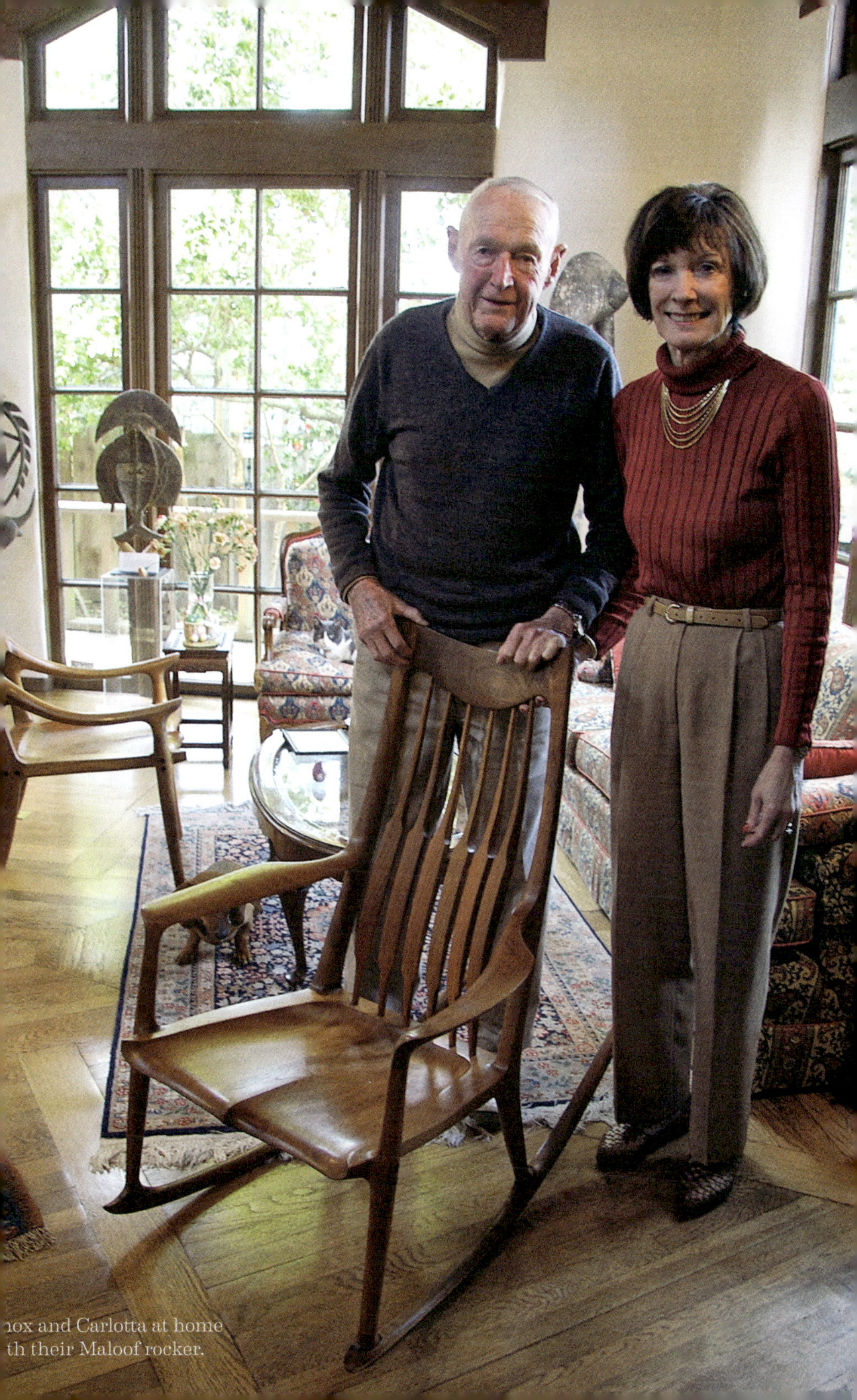

nox and Carlotta at home
th their Maloof rocker.

know Millard Sheets, the department's charismatic head, and his daughter, Carolyn, who attended nearby Girls Collegiate School where Knox would soon teach ancient history.

"No question—it all began with Millard," said Knox. "Millard had a following of young, aspiring artists and craftspeople, and Sam got to know most of them. And Sam was not shy. Among the artists, he was in his element."

Knox's smile blazed at the recollection. The sun slivered through his living room window, landing at the top of each chair.

"I'm eighty-nine years old," he said, "and my life has been motivated and influenced greatly by people I've met. Certainly, I can say that about Sam. As we got to know each other, things that he would do and say had an effect on me. I wasn't knowledgeable about craft before I met Sam. He changed my vision. You learned so much if you were around him."

In the early 1970s, Carlotta was teaching history at Immaculate Heart College in Los Angeles, a Catholic women's institution run by the order of the Immaculate Heart of Mary—"a marvelous college," she remembered. She shared an office with a brilliant political scientist who was also a nun in habit and a friend of Jerry Brown, then secretary of state. Long interested in public service, Carlotta joined in their conversations—a sparkling mix of high abstraction and practical politics.

In 1972, Knox was also teaching at Immaculate Heart College, and met Jerry Brown through a colleague. Knox decided to abandon his sabbatical year, postponing research and writing to join Brown's campaign for governor as director for the statewide organization of volunteers. Carlotta also joined the campaign, working on women's issues and speaking enthusiastically throughout Orange County on behalf of the candidate.

When Brown won, he picked Carlotta as his appointments secretary. Her duties involved the selection and vetting of two thousand political appointments, enabling the new governor to redeem his pledge to usher into state government historic numbers of women and people of color.

"Carlotta's own job had always belonged to a sixty-year-old man," recalled Knox. "When she arrived, their faces dropped."

With Carlotta ensconced in Sacramento, the Brown administration searched for a slot to suit Knox. Though new to government, he was not a political novice. In 1962, Knox ran for Congress in the Republican stronghold of Pasadena as a Democratic "peace candidate"—in the context of the times, an internationalist opposed to Cold War policies.

"In 1966," said Knox, "Congress passed an earthshaking piece of legislation, the National Historic Preservation Act. Its objective was to save historical properties across the country. It established in every state an office of historic preservation, and asked the governors to identify a qualified individual to serve as head."

In 1977, Knox became California's first state historic preservation officer, or SHPO, serving eventually under Governors Brown, Deukmejian, and Davis. Over the years, Knox became a respected expert in the field, later working as a private consultant on numerous historic buildings, including sites at UC Berkeley, UCLA, and the Downtown Central Los Angeles Library. "I learned all my historical preservation on the job," he said.

Key to the legislation's influence was Section 106, empowering the SHPO to devise strategies for mitigating adverse impacts on historic properties. That included properties threatened by projects supported to any degree by federal funds.

Such as freeways.

In 1990, the California legislature at last authorized construction funds for the long-rumored, perpetually delayed twenty-eight-mile extension of the Los Angeles Foothill Freeway.

"Sam was crushed," remembered Knox. "The freeway was going to destroy his house and workshop. That's when I came back to head the Maloof Foundation and steer them as to how to plan and proceed. And how to pacify Sam, who was getting irritable, desperate, and depressed."

Since Knox was no longer serving in state government, he could apply his expertise and experience—and his contacts in Sacramento—to the complicated task of preserving the original spawning ground of the American artist he knew best. "What were the alternatives?" asked Knox. "Go under, over, around?" Mitigation turned out to be relocating Sam's house and workshop several miles from the original site.

The California Department of Transportation (Caltrans) initially resisted the move. "They were not what you'd call a friendly neighbor," said Knox. But after four years of complex discussions, the relocation was approved and funded. From the historic preservation point of view, the massive project—California's most ambitious to date—was regarded as a model. Caltrans and San Bernardino Associated Governments (SANBAG) got the freeway they wanted. Preservationists saved the buildings. The public had access to an artistic treasure. And for Knox, the extended effort "renewed my friendship with Sam."

Still, Sam was "not at all happy about relocating," remembered Knox. "He went into the move kicking and screaming."

"It was hard for him to give up ownership and have people coming into his house," said Carlotta. "Sam and Alfreda had been very generous on their own in letting people see the original property. But now there were tours on Thursdays and Saturdays. He still felt it was his house and that he could do or not do whatever he wanted. He got upset when a neighbor complained about too many people on site."

A discomfiting, almost surreal scenario: The master craftsman finds that the home he has meticulously, exquisitely designed and built over the decades and the workplace where he has labored most of his life have been transformed into a public institution. Some days, it seems like he has become an institution. An object of admiration, certainly, but also of curiosity and diversion.

Carlotta joined the Maloof Foundation as executive director in January 2002, setting up the bookstore and crafting tour notes for the docents, along with descriptions of Sam's furniture and the work of the many artists he collected. Throughout the week, she and Sam had countless casual contacts, along with more substantive conversations detailing aspects of his life and art that found their way into the docents' lectures. Up close, she recognized his dissatisfactions, the irritations. He groused about the noisome exit signs placed above the doorways, *his* doorways, a fire code stipulation that underscored the fact his former home now actually belonged to the foundation. But Carlotta also paid attention to more subtle, reemergent qualities of the man's character. His abiding interest and concern for people, despite the present resentments. The occasional feelers of engagement sprawling like tendrils from his workshop and out into the grounds and garden now trodden upon by strangers. And on some days, it was the accommodations that seemed paramount, the familiar courtesy and the enthusiasm she and Knox had associated with their friend over the decades. Sam would join a tour group walking through the house, and he'd start to explain something about a hinge he devised for a cabinet door. Or the redwood latch on the front gate. Or the inspiration for a table and set of chairs. Or the qualities of ziricote or cherry or English brown oak or walnut.

And then he'd invite these visitors, these strangers—his guests—into the workshop, which was supposedly off-limits. Sometimes even into the new residence he had designed.

Just like that, he'd invite them in.

21.

NEIGHBORS
JOHN AND MARILYN HOCKING

Years ago, while still a student at Wisconsin's Beloit College, Marilyn Hocking decided to spend a weekend with friends in Chicago and devote an afternoon to exploring the Art Institute. Wandering through the galleries, she spied an exhibit featuring a gorgeous rocking chair. A plush cord encircled the chair. Marilyn casually surveyed the gallery. Nobody was looking.

"So I sneaked behind the rope," she recalled, "and touched it."

The feel of the wood, its shape and smoothness—the irresistible temptation.

An invitation, really. She would never forget it.

Toward the end of Marilyn's senior year, the recruiters descended on Beloit. California was now the most populous state in the union, desperately in need of teachers. The recruiter for Pomona Valley's public schools flashed a fan of color photographs depicting orange and lemon groves—a sly provocation in the Wisconsin winter. The background featured the San Gabriel Mountains with a snow-capped Mount Baldy characteristically vivid on a windswept afternoon. The coast proved a short drive, the temperatures were perfect, and the recruiter nonchalantly recalled a friend who was almost certainly deep-sea fishing that afternoon.

You need to move out here, he told her.

Marilyn started her three-decades-long education career at Pomona High School. One afternoon, the football coach mentioned that he had to attend a ceremony for his uncle, a furniture maker, who was receiving an award of some kind.

"He's not Sam Maloof?" Marilyn asked.

"I hadn't a clue that Sam lived in the area," she recalled. "But I loved his work and I had remembered his name."

When Marilyn and her husband, John, moved to Alta Loma, Sam and Alfreda turned out to be their neighbors. Over time, they became friends. Not the "inner circle," stipulated John. Rather, their relationship fell somewhere between intimacy and supple acquaintance, nurtured by time and opportunity.

"I want you and Marilyn to come on up," John remembered Sam periodically insisting. "We've got some furniture here you haven't seen; you got to see this new stuff I'm doing."

"He'd show us around the house," remembered Marilyn, "the old parts and the new pieces. The furniture he was working on and how he was slightly changing the design. It was so much fun. He and I started talking politics, and we'd have a great time." Eventually, Alfreda would materialize at the shop door and suggest to Sam that the hour to get back to work had arrived. John and Marilyn would return home until the next invitation or spontaneous visit or convivial summons to join a bevy of family and friends for one of the huge birthday parties, with In-N-Out Burger shoveling hot meals off the truck.

They also started to buy furniture: A rocking chair that somebody had ordered, but failed to pick up—they could have it for $650, a fortune back then. The dining room table and six chairs, the center of Hocking family holidays for many years. A rocker for Marilyn's sister, another for her brother. By then, Sam was world famous, his furniture acquired by museums, celebrities, and CEOs. Marilyn taught math and then went into school administration. John taught English and headed up Chaffey High School's academic decathlon. Sam might have been a celebrated artist, but he was still their neighbor, an ordinary person in the best sense of the word who happened to be blessed with extravagant talent and an unflappable resolve to probe its boundaries.

John ordered a bright rosewood desk, shaded in blacks and reds. He'd seen a similar piece Sam made for himself during a visit to the house.

"I wanted drawers for my papers. I had six classes then, so I wanted six drawers. I wanted something to write on so I could comfortably grade papers. Sam designed my desk based on the piece in his house. Then he scaled it down. When I tried to put 8 1/2 by 11 paper in the drawers, they were too small." John shrugged at the recollection of his plans colliding with Sam's vision. "They looked better smaller," he admitted.

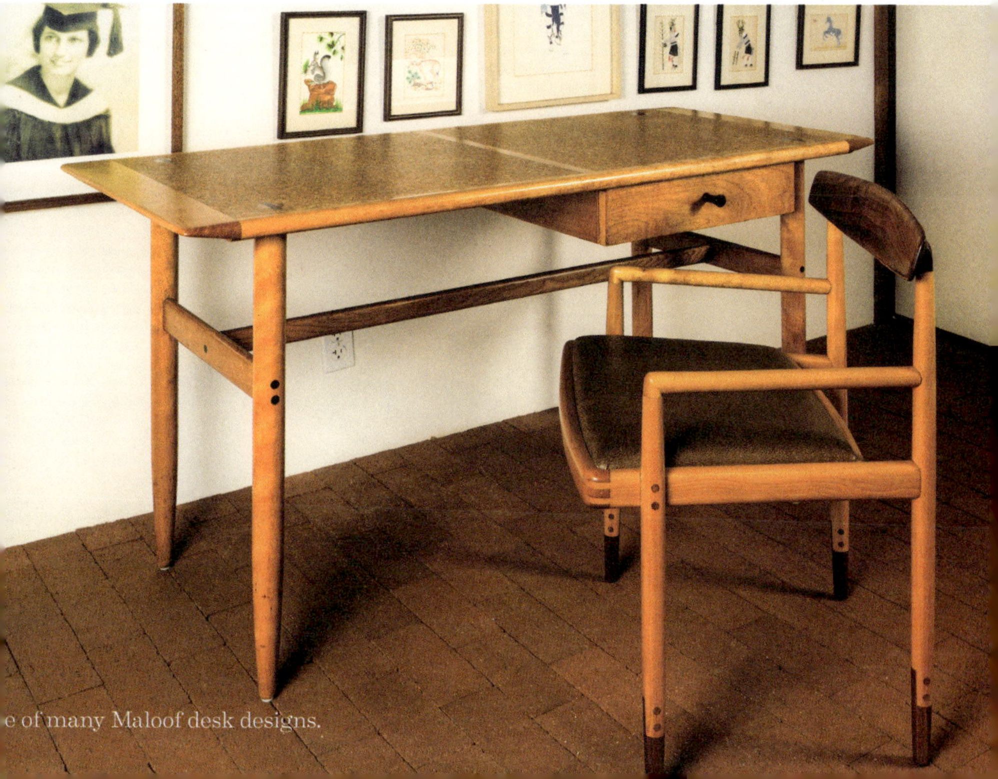

e of many Maloof desk designs.

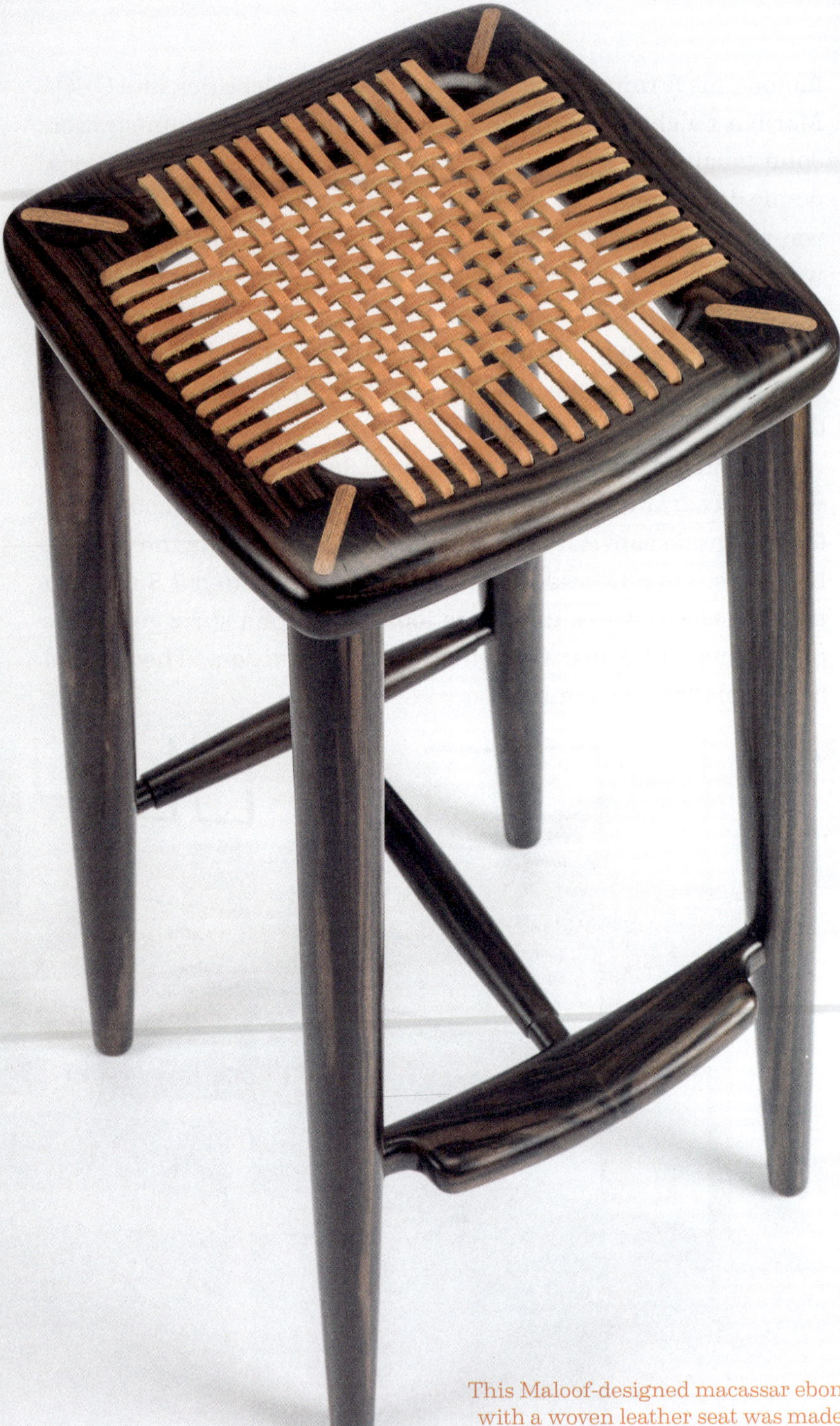

This Maloof-designed macassar ebony st
with a woven leather seat was made in
Maloof Woodworker shop after Sam's passi

John and Marilyn also tried to commission a pair of barstools. Sam completed the design and they signed a contract.

"We never did get them," admitted John. "After nine or ten years, Alfreda died. I said, 'I'm not going to bug Sam.' He was really having problems. 'Let's just forget about it.'"

After eleven years, John called the shop and cancelled the order. Not with rancor or exasperation, but in amity, goodwill. Because that's what people do when their friends and neighbors find themselves in trouble.

Early in their acquaintance, Marilyn had told Sam about her youthful visit to the Chicago Art Institute and her encounter with the rope and the chair. Sam loved that story. It made a perfect counterpoint to one of the tales told most often about Sam—a parable frequently employed to convey an essence of the man whom countless people felt they knew and understood, despite having met him perhaps only once or twice.

Sam was roving through some stately museum. (In the many enriched retellings, the location varies, shifting from the Smithsonian to the Getty or the Boston Fine Arts Museum, and who knows where else?) Sam turned a corner and suddenly came upon one of his own rockers cordoned off with signs admonishing patrons not to dare touch. Naturally, he dismantled the barriers and took a seat.

Now consider the craftsman—momentarily indifferent to his standing as an artist. Insistent in his regard for use as much as beauty. And entertain the possibility that his vision and commitment might originate with a desire to refashion the world for the delight of his neighbors and friends.

22.

MARRYING SAM
CAROLYN OWEN-TOWLE

"Please," said Carolyn Owen-Towle, "take a seat. Once people are seated, they don't want to get up."

And, yes, it's true. The astonishing pleasure of simply sitting. The chair seeming to conform to the body, while the body gradually locates its contours and supports. An accommodation between wood, flesh, and bone that feels essential to any visceral encounter with Sam Maloof's furniture.

"There's a sensuous invitation in the shape of the pieces," said Carolyn. She points to a far older piece of furniture in the corner of her San Diego living room. "I really love that chair too. It came around the Horn of Africa with my grandfather when he arrived in California. I love it for its history, for its presence in my grandmother's home when I was growing up. I have very different feelings about Sam's chairs—feelings unlike what I experience with any other furniture. Partly because I knew and loved him."

Actually, Sam knew Carolyn long before she was aware of his existence.

"The first time he showed up at our house in Claremont," she explained, "my mother answered the door. I was a baby in her arms, crying about whatever babies cry about. I don't remember seeing him again until I was eleven and Dad hired him to make a run of silk screens. That's one of them on the wall."

A nod to a sixty-year-old print of a Mexican *campesino* created by Carolyn's father, Millard Sheets, Sam's early mentor.

"Sam was part of our family," remembered Carolyn. "He worked for my father for several years, helping to construct buildings on the grounds." Together, they built walls, a bird sanctuary, a barbecue station. "Sam did so many things. He made signs—his script, his handwriting, was beautiful. Exquisite. For more than a year, Sam even lived at our house with his bed in Dad's studio. He hadn't started making furniture yet. But Dad drew Sam into the art world, introducing him to so much that he had never experienced before. They shared this enormous interest and excitement about art. About people in the art world."

Both recently returned from the war—Sam had been stationed in the Aleutian Islands, Millard in India and Burma, where he served as an artist-correspondent for *Life* magazine and the United States Army Air Forces—they forged their relationship around an insistent appreciation of beauty, an antidote to the horrors that had so recently verged upon their lives. "Dad was a collector," remembered Carolyn. "Sam became a collector."

For Carolyn, the proximity to the open-hearted "gorgeous" young man proved intoxicating.

"I always had a yummy relationship with Sam," she recalled. "We just loved each other. He was so engaged with life. If you spent any time together, you became a friend. A genuine one. Over the course of a year, I developed quite a crush on him. My first. A schoolgirl crush is a very sweet thing. No actual romance, no sexual connotations. It's just, *Ah...this beautiful man....*" The two of them talked for hours about every imaginable subject, Sam now opening up new worlds to Carolyn. He gently teased her that when she grew up, they would marry.

"Sam was kind," said Carolyn, "proper. He'd go out at night with various women, and I'd tiptoe down to his bed, turn down the blankets and sheets, and put a camellia on his pillow. I remember one time as a child when he took me and a young woman to the Hollywood Bowl. It was like a date with Sam—for us both. Sam even let me walk between them. I should blush to think of it—though these days, to tell the truth, I can't blush any longer."

When Alfreda entered Sam's life, it struck Carolyn as an immense, capsizing surprise.

"One night Mom and Dad were having a big party in the living room. I was in my room minding my own business when I heard Dad say, 'Go ahead, Sammy, kiss the bride!' I got hysterical. I cried my eyes out. Mother and Dad came into my room and sat down on the bed. They told me, 'Someday you're going to be so ashamed of this.' And I thought: *Never!*"

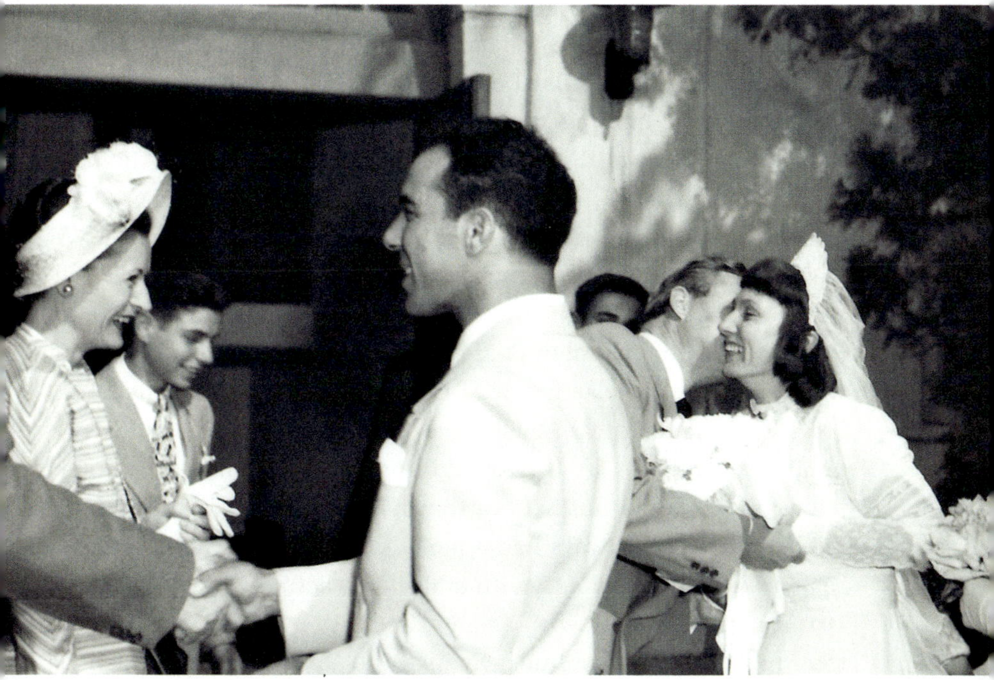

Millard and Mary Sheets, left, offer wedding congratulations to Sam and Alfreda.

When Sam and Alfreda married, Millard gave them one of his paintings. A Liberty ship sailing out of safe harbor and into the Pacific war, a very dark and brooding picture—a reminder of the bloody world they had both left behind and a paradoxical invitation to the peaceful, industrious haven being established in the Pomona Valley.

As a model for marital life capable of sustaining a serious artist, Sam needed only turn to Millard and Mary. The couple loved each other deeply—"passionately," recalled Carolyn—and worked closely as partners, though Millard's career took precedence. Mary shared her husband's profuse creativity. She was intelligent, expressive, verbally adept, ever curious. Carolyn adored her parents. Even as a child, she appreciated their range of friends, from laborers tending the town's wide, tree-lined streets to the great violinist Jascha Heifetz, whose son attended school with Carolyn's brother.

"Claremont was interesting, rich, stimulating, beautiful," remembered Carolyn. "People came early in the century and built marvelous homes." Millard designed a striking modernist home, the subject of a photo essay in *House Beautiful* that also drew the interest of editors at *Life* then eager to portray the look of the new as California shouldered its way into national consciousness. "My father taught my brothers to lay bricks and work in the garden. We were all instructed that nothing was worth doing unless you did it well." Millard painted in his home studio with ferocious energy, though as a child Carolyn paid scant attention. "I knew my father was important," she said. "But I didn't think a lot about his fame. He seemed indefatigable. He used up his life, and that's exactly what he wanted to do. If he could imagine something he wanted to create, he could not believe that he couldn't finally do it. He'd figure it out. Dad always had a pencil in his hands. Woe be to the restaurant," she laughed, "with tablecloths actually made of cloth."

Carolyn flipped slowly through her father's sketchbook. Page after page after page of line drawings and shaded figures, the models captured on the fly at home or in faculty meetings or conjured from imagination: the irresistible pulse of Millard's prolific hand generating rabbits, horses, caricatures of colleagues. A portrait of young Carolyn pouting. Another of her brother slouching. "These drawings were just lying around when we were children," she recalled. "Mother gathered them together and I put them in a book. Seeing is something our dad gave us all, an incredible gift. To see, to understand, to appreciate. To this day, I'll get stopped by something that's beautiful, and I just have to comment. Dad would do the same thing. 'Say, look at the way that tree is shaped.'"

While growing up among artists enriched Carolyn's life, she acknowledges the complications.

"From the time you're small," she said, "people pat you on the head and ask, 'Are you artistic, too?' You just want to climb in a hole."

Yet when she did find her calling—Carolyn is an ordained Unitarian Universalist minister—she recognized the affinities with her background. The drive to add to the world's beauty, she believes, is spiritually attuned, the mystery of creation endlessly echoed in works of art. For many intensely creative people, including her father, the perception of a spiritual basis underlies their vision, quickens their invention.

In her tender memoir of Millard Sheets, *Damn Gorgeous,* Carolyn remembered her father telling her "that he knew a power had created this miraculous universe. Surely, this power was the animating spirit of a beautiful and harmonious world. Nothing could be this magnificent without some plan. He hadn't the slightest doubt about it. Millard Sheets never felt he was alone in his creativity. There was a greater force guiding his creative moves."

The force gripped Carolyn too, though shifting her in a decidedly different direction.

"At age thirty," she recalled, "I recognized that religion was a huge part of this universe, and I knew very little about it. I originally went to church because of the obligation I felt toward my children, but I ended up being the one who was fed. The church brought out my intellectual, joyful, practical, problem-solving sides. People would confide in me—I had a natural ability. One day, I just recognized: 'I think I'm a minister. I guess I should become one.'"

She devoted four years to an independent study program "intended for women like myself who couldn't take off from their families." Women still accounted for a small percentage of the clergy. In the spirit of the times, Carolyn stepped forward to assume leadership: first woman president of the Unitarian Universalist's international service committee; first woman co-minister (along with her husband, Tom) of a large metropolitan church. When she ran for the presidency of the denomination, she lost by only a handful of votes.

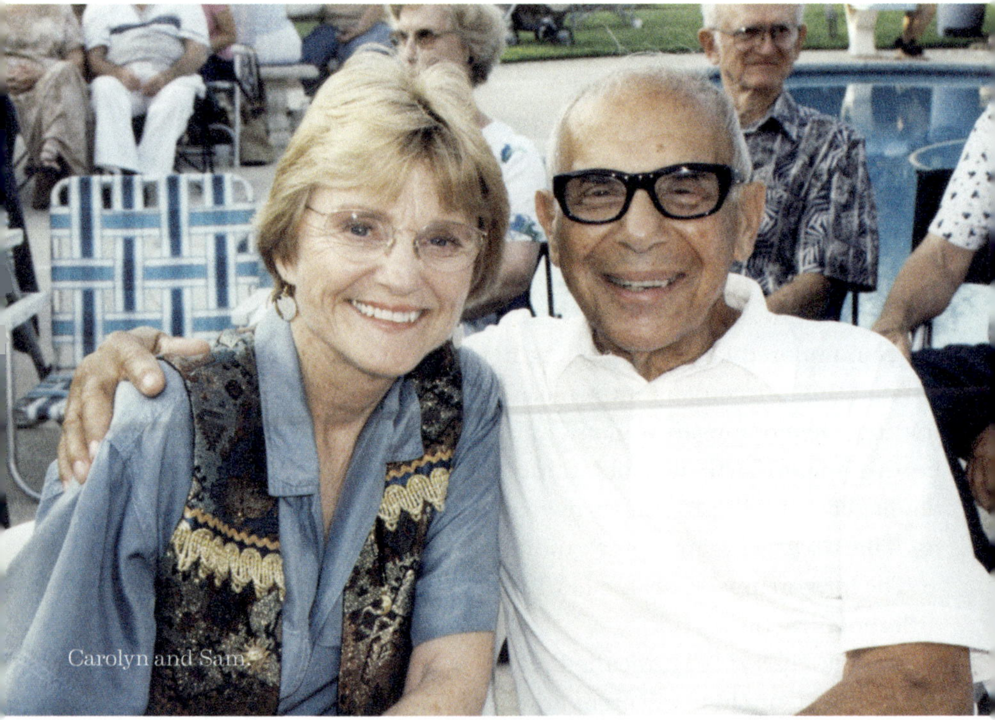

Carolyn and Sam.

"Sam was also very much a religious man," she said. "I'm spiritually moved by Sammy's work. 'When you do something with wood,' he used to say, 'do something the wood likes.' If you were a tree, what could be better than to become something so lovely and exquisitely made? Just think. A tree is God's creation, and out of that beautiful life comes another life as one of Sam's creations."

Carolyn's calling as a minister also led to the fulfillment, of sorts, of an unlikely childhood aspiration. Finally, she married Sam—to Beverly Wingate, one of her closest friends, in a service she officiated along with her own husband.

"Sam had grieved incredibly over the loss of Freda," Carolyn recalled. "We were afraid he'd go, too."

When Sam and Beverly got together, it closed a circle of acquaintance, friendship, and family spanning a half century.

"Years ago, I had moved to Riverside with my first husband," said Carolyn. Sam introduced her to Beverly, who also lived in the area and had purchased some of Sam's furniture. "We bonded and became best friends. We raised our kids together, went to Sea Ranch together where we all bought homes. We've been friends for more than fifty years."

One day, Carolyn received a phone call from Bev. "'What are you doing Sunday?' she asked. 'I'm preaching,' I told her, 'but I'd love to see you after.' She told me that she had a new friend and planned to attend the service. Tom and I lay in bed, wondering who could it be. Bev was so urgent about seeing us. So that morning, we're in church, Tom and I sitting on opposite sides of the altar. And here comes Beverly walking in, holding hands with Sam."

23.

HUG
BEVERLY MALOOF

All month, Bev Maloof has been wandering through the house, rearranging flowers, allowing her gaze to settle on a painting or pot, her footfalls tracking through the outside gardens where she surveys the yarrow and sage, the eucalyptus windrow, the stone statuary and sculpture. *Soaking it all up,* she tells herself. Six years after Sam's death, Bev is taking some time off from her usually busy schedule—a month of porous receptivity to the world she inhabited with Sam in the final years of his life. A consolation in his absence.

"I've hardly accomplished anything," she said. "I don't know how to explain it." The day before, she managed to straighten out her desk. Then it was back to heeding the artwork, allowing a flinty strike of memory to spark some association with the artist or a recollection of a studio visit or some anecdote that Sam had related, probably on many occasions. "I've been here for a long time," she admitted. "You'd think it'd be all soaked, all soaked."

Not thoughts, exactly. More often, a whisper of some moment's feeling tone. When she halts before the earthy glaze of a McIntosh ceramic, the tan or cream or obsidian vessel curved like the surface of an egg, rounded like a gourd or shoulder or waist, her attention filters back to the artists' lunches at Walter's in downtown Claremont. Harrison McIntosh joining Rupert Deese and Jim Hueter on the restaurant's front porch as the summer breeze cools Yale Street and Sam holds forth about whatever was exciting Sam at the moment. The four friends shouldered together in a ruffle of enthusiasm, their conversation secured by Sam's energy, his stubborn youthfulness and stability.

"I'm giving myself a vacation," Bev decided. A month's respite from her typically endless lists of tasks so that she may lavishly absorb all aspects of her happy life with her second husband, the artist she had admired for decades before ever imagining that she might marry him. "I've been looking at the trees. When we moved in, it was pretty much a bare lot. They had just moved the trees from the other site. The palm tree," she stipulated. "Sam fussed and fussed, saying it was going to fall over. Now it's huge, huge. And it's not going to fall over."

Today, a dozen New Zealand tourists streamed through the original house and studio while Bev maintained a polite distance. She's been visiting the old house during quiet hours—the home of Sam and Alfreda, friends she and her first husband made a dreamy half century ago. Her own private tours of the house, roving between rooms, admiring the tree trunk post supporting a ceiling beam or fondling a walnut latch designed and sculpted by Sam, sometimes brings back events she has only heard about, key moments in Sam's life that warranted his persistent review and reflection.

The first Conference of American Craftsmen, held in 1957, at Asilomar on the Monterey coast. More than 450 people attending. Crafts professionals and advocates expressing concern about their role in twentieth-century America as machine production threatened their ethos and livelihood. Sam liked to tell the story about a renowned artist—*what was his name now?*—who responded with great urgency when Sam asserted the value of handcrafted versus machine-made furniture. The artist, Wharton Esherick—*that's who it was*—a brilliant woodworker and sculptor, thumped Sam on the chest and ordered him to "Never change your mind." Sam held fast to the occasion for years, confirmation that he was on the right track.

Wharton Esherick. Seventy years old at the time. His fifty-year career spanned styles from Arts and Crafts to expressionism to the curvilinear free-forms that made his reputation; his rural Pennsylvania home and studio, designed and crafted like Sam's, is now also a museum filled with his own art. Sam knew Esherick's house from magazine photographs, took them as a source of inspiration for his own home.

"Asilomar is how they started meeting each other," recalled Bev, "these artists from all over the country. After I married Sam, we thought it would be fun to get them back together, and we did it. One woman even came from France."

Beverley Wingate first encountered Sam Maloof's furniture at the *Arts of Daily Living* exhibition as part of the Los Angeles County Fair in 1954. Her mother, a lifelong art student, had enticed Bev and her first husband to attend so that they all might consider a piece of sculpture contributed by her current teacher.

"I walked into this room," said Bev, "and it was furnished like a den. There was the most beautiful table I'd ever seen. Beautiful proportions, beautiful grain. I cried. It was so beautiful. I cried." Her husband was also moved and suggested that they contact the table's maker to see if he could fashion one for them.

"So we called on Sam," recalled Bev, "and told him that we were newly married, didn't have furniture, didn't have a house. But someday, we'd love to have that table. Could we send a little money each month? I sent thirty-five dollars every month until we had our first house. Then I called Sam. 'We bought a house. Quick! Make the table.'"

Three years later, Sam notified them that the table was finished.

"We never got Sam's chairs for the table. We ended up with some Thonet chairs I found somewhere. Partly because Sam's were too expensive. Partly because we needed something to sit on."

Over the years, Bev and her husband visited Sam and Alfreda, attended events at their house, their acquaintance ripening into friendship. "We'd buy things periodically. Sometimes they had a show, and we'd go to that. Sam loved to talk and my husband loved to talk. He'd say, 'Let's go over and see Sam.'" The men convened in the shop, Sam taking a breather from shaping or sanding. Eventually, Alfreda would break away from her conversation in the house with Bev and step out to invade her husband's territory, affably reminding him, "Time to get back to work, Sam."

In the spirit of midcentury California, everybody from someplace else and endeavoring to fashion a convivial roost within the madly expanding state, the young couples articulated through their spontaneity and zip a sense of possibility that nearly served as a moral tone. California had heralded a promise for daily living, the countless aesthetic choices of several million new households adding up to a purposeful style, from tiki lamps on Moorish patios to ranch house homemade barbeque pits to the assiduous, acutely discerning regional modernism then being handcrafted on a daily basis by Sam Maloof and his peers.

Decades later, in the late 1990s, Bev would meet Sam in starkly different circumstances. Alfreda had recently died. The move from the original site to the new house, studio, and grounds some two miles away had begun, with Sam feeling more besieged than relieved—and certainly not yet settled and appreciative. Friends worried about his state of mind, his loneliness. Bev's close friend, Carolyn Owen-Towle—the daughter of Sam's early mentor, Millard Sheets, "practically my sister," Bev affirmed—kept her apprised of Sam's projects and difficulties. But they did not decisively meet again until years after Bev's divorce when a colleague in her real estate office suggested one afternoon that they attend an art sale

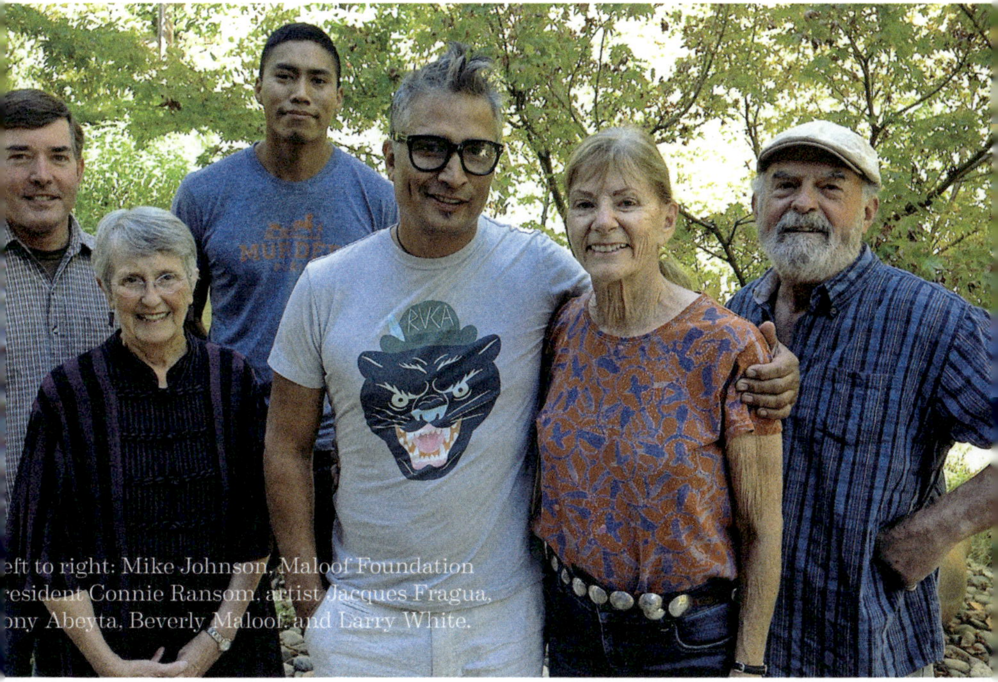

eft to right: Mike Johnson, Maloof Foundation
resident Connie Ransom, artist Jacques Fragua,
ony Abeyta, Beverly Maloof, and Larry White.

in Claremont. On the drive from Riverside, Bev talked about Sam. When they arrived at the gallery, he was standing on the porch.

Bev rushed up the stairs and wrapped her arms around in him in a warm, sociable hug.

"I wasn't thinking of anything but friendship," she recalled.

But sometimes, as with a glance or a rare brush of the hand or a trusted friend's spontaneous embrace, the terms of a venerable alliance abruptly shift. The never-before-spoken possibilities crystallize over the course of a barely perceptible instant into aspects unsought and unexpected.

"I think that must have happened with Sam," confided Bev. "When Sam made up his mind about things, he really pursued them. Later, he told me, 'That hug felt so good.'" The courtship proceeded swiftly. Sam called with invitations to lunch. Circumstances seemed to demand his presence in Riverside. "Things just happened that threw us together," said Bev.

They shared a delight in viewing sculpture and painting, contemplating and discussing the artist's skill, discipline, intuition. The world, if you cared to look, offered an astonishing variety of pattern and harmony. The beautiful world enticed and surprised. Bev proved every bit as eager as Sam to soak it all up.

Following her divorce, Bev had spent more than twenty years as a real estate agent, benefiting considerably from her ability to enter the jacketed, sometimes obscure sensibilities of her clients, perceiving in precise terms—the ocher walls, the angle of a backyard slope—exactly what they desired when they did not quite know themselves. An art lover's instinct. The heightened perception borne of a zeal for looking and a condition of feeling that underlies the most acute vision.

Then Bev entered Sam's world. "He wanted that relationship. And I did too. It surprised me. But it felt good for me, and it felt good for Sam." They married in 2001.

Was it at all difficult being married to a famous artist?

"No. Not that famous artist. Sam was not easy. He would get angry, he wasn't a perfect person. But I really, truly loved him. I understood him. When he got upset, I knew it was just temporary. He was a complicated person because he had so many parts. He had sensitive parts, visual parts, he had weaknesses based on life's experiences. But I loved him, I really loved him."

The phone rang. Carolyn was driving up from San Diego, but the freeway was clogged; she was running behind. They'd eat together later that evening.

"You have boyfriends," explained Bev, "you get married. Later you have friends who you do things with. I was born in 1931. So how old was I when Sam and I married? I was seventy! I couldn't believe it. I said to myself, 'I'm in love!' I'd been on my own for years, perfectly happy. My real estate business, my community activities, lots and lots of friends. But oh, boy—I gave it all up for Sam."

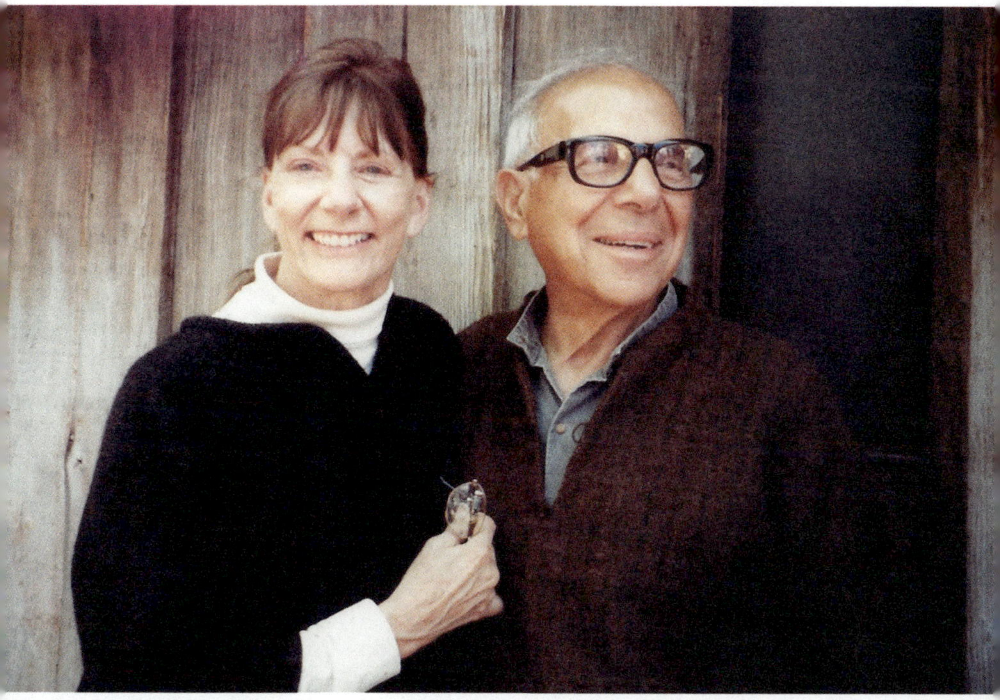

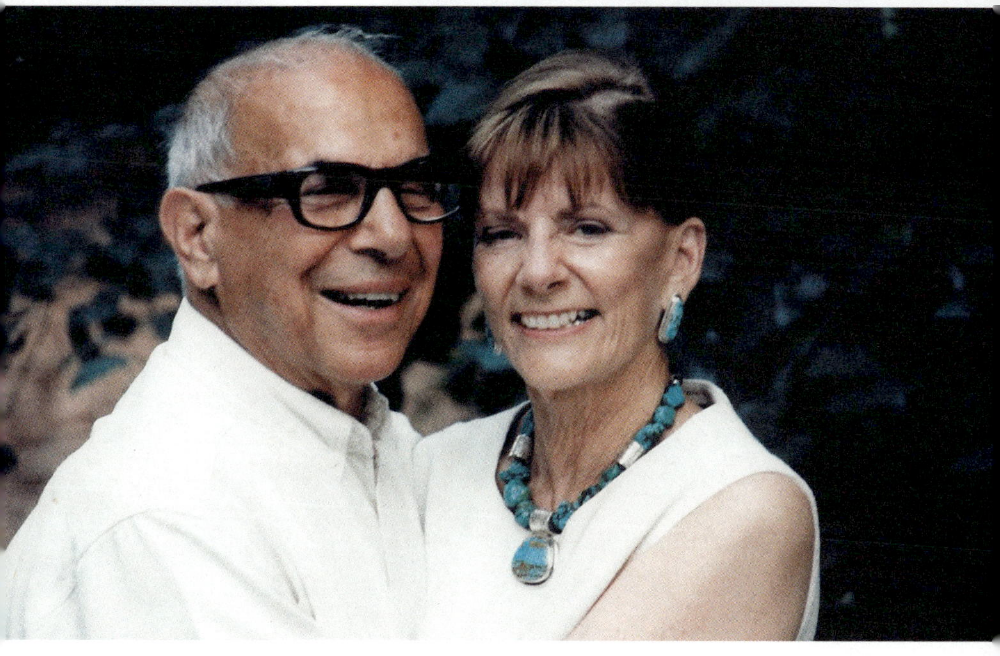

24.

SAM IN THE WORLD

In 1963, Sam journeyed to El Salvador to spend three months tutoring local woodworkers, a State Department project aimed at bolstering the regional economy—one of the more temperate strategies of the Cold War. "The people were wonderful to work with," Sam wrote twenty years later in the midst of the Central American nation's horrific civil strife. "How sad the turmoil there now." As always, Sam had cultivated warm friendships with his colleagues, their conversations conducted in Spanish. Growing up in Chino, California, a dozen miles from Alta Loma, Sam had picked up the language before learning English.

Beirut, Lebanon. En route home from a similar mission in Iran, Sam wandered into several woodworking shops, accompanied by an interpreter who persistently misrepresented his interests. Referred to only as "that American," Sam was alleged to be a US government official capable of ordering thousands of dollars' worth of furniture from the city's craftsmen.

"I do not think you are telling them what I asked you to tell them," Sam interrupted.

"Oh, yes, I am," replied his interpreter.

So Sam responded for the first time that day in Arabic, another linguistic boon of his childhood.

Sam Maloof—the roving, multilingual, cosmopolitan artist.

In truth, Sam managed to effectively negotiate the world without acquiring a veneer of worldliness. Rather than the sophisticate's pose, he often displayed a joyous innocence, a willful naiveté.

A close associate recalls an occasion when Sam had been invited to a luncheon in a private home following one of his lectures. Along the way, he chatted up some admirers and blithely invited them to tag along—a discomfiting surprise for his hosts. Sam tended to interpret social conventions and civic requirements by the light of his own values. When designing and building his house, he seldom granted primacy to permits or codes or what he perceived to be the whimsical preferences of officialdom. The world would shake itself out accordingly.

"At times, he made it sound like he was fabulously wealthy," said another friend. A compensation, perhaps, for growing up in a struggling immigrant family. Yet, "Sam was rarely frivolous," emphasized another intimate. He never indulged in snobbery and could gently distance himself from an affiliation that he didn't genuinely prize. When Sam was recruited to present a rocking chair that had been purchased by a supporter for President Reagan, he amiably complied, but shied from basking in the reflected glory. "How was it?" asked a friend. "Weren't you just amazed at being in the White House?" "Well," said Sam, "the Oval Office is very impressive." "Yeah. What about the president?" "The Oval Office," he responded, "is very impressive."

Sam cared little for vacations, though in the early years, he did take his family almost every summer to Arizona and New Mexico. They would visit the pueblos where Alfreda had taught in the Indian Service, and she was hailed as a returning hero, laden with gifts of jewelry. As Sam's reputation and means increased, the family might have toured widely, even luxuriously. But Sam preferred to yoke his travels to his work.

Over the years, Sam participated in over one hundred art exhibitions. He received honorary degrees, including a doctorate from the Rhode Island School of Design. He gave scores of lectures across the country and abroad, even speaking at the Vatican. He conducted scores more workshops with numerous recurring appearances at the Anderson Ranch Arts Center in Aspen, Colorado, and Highland Woodworking in Atlanta, Georgia. In Washington, DC, he delighted in the anonymity of the big city, the wall posters publicizing his Smithsonian retrospective flitting past as the underground Metro delivered him to the galleries—Sam the secret sharer in his own burgeoning fame.

Wherever he traveled, Sam carried with him a portable sense of security—especially on the occasions when Alfreda could join him. He mixed freely with strangers, his unstudied poise seemingly at odds with his standing as a major artist. As he came to be celebrated as a master craftsman, the preeminent woodworker of his generation, his projection of modesty and simplicity became an essential aspect of the public image. "I'm sure you could talk to hundreds of successful woodworkers who would say Sam was a big influence," said a young furniture maker, "whether or not they ever met him. The idea of him. The essence."

As Sam aged and pared back on his work-related travels, his furniture served as his primary emissaries into the wider world—even as his admirers increasingly flocked to his home. The house, shop, and grounds all spoke to the advantages of fashioning a life according to one's own unfolding, exacting specifications. The rewards of remaining vigorously in place.

To the end, Sam was an effortlessly public man—not a trace of the hermit, despite his marvelous hermitage. He set everyone at ease because he seemed so unruffled himself. Sam almost never wore a jacket and tie. Even when he received the Medal of Honor at Ellis Island, he shrugged off the strictures of formality and slipped into one of his knit sweaters. They pinned a boutonniere on his chest. Everybody agreed that he looked just fine.

25.

THE BUSINESS
ROS BOCK

It seemed a very odd interview.

"What would be my duties?" Ros Bock asked reasonably enough.

Sam shrugged a reply.

Whatever needs doing, he appeared to be saying. *Anything you want to do.*

"And the hours?"

Another shrug.

After seventeen years spent working for a construction company, weathering Chapter 11 and Chapter 7 bankruptcies and enduring a daily hundred-mile round-trip commute, Ros felt ready for a change. So when word reached her that Sam Maloof, whom she had met briefly more than twenty years before, needed someone to help with his business, she expressed a tentative interest.

What did Sam think he needed? Someone to answer the phones, certainly. Keep track of incoming requests, outgoing orders. Manage the checkbook, pay bills. Keep the records and files straight.

Do what Alfreda did. Whatever that might have been. Another assistant had helped around the office toward the end, but the job had always been defined, embodied really, by Alfreda. Two years had passed since Alfreda's death. Sam referred always to her "passing." September 23, 1998.

He looked aggrieved and gloomy each month when the twenty-third rolled around. Not just in September, but every month.

For reasons Ros couldn't quite identify, the job felt right. It wasn't because she had cultivated a profound interest in art. Her knowledge of the subject, she liked to say, extended no further than knowing not to wear stripes with plaid. She had already reached the age where most people contemplate retirement. She would need to take a pay cut.

She accepted.

"Sam was a very poor businessman," said Ros, summing up what became evident almost immediately. "A great artist. But his business sense...well, it just wasn't important to him."

Did he have money in the bank? No worries then. If the business needed a little extra to cover payroll, he could always dip into his personal account. Was the operation profitable? He couldn't say

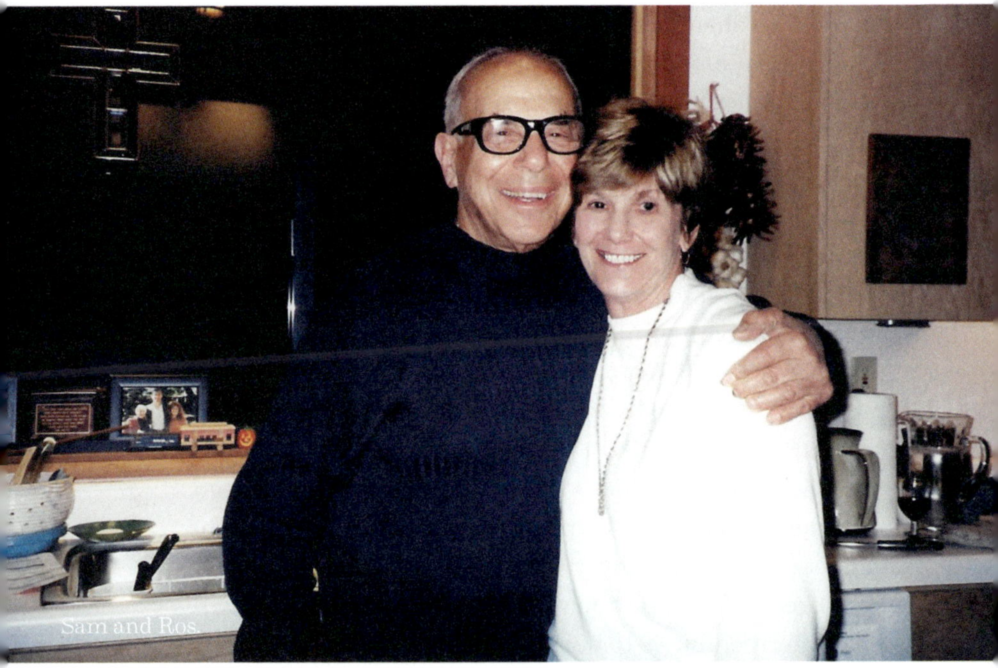

Sam and Ros.

precisely—not until Ros began running the numbers on QuickBooks to provide a monthly profit and loss statement. Basics still required attention. Authorizing checks to be signed on the corporate account instead of the trust. Filing tax information under the proper identification number. Ros had to accustom herself to covering for Sam when he took his workers out for lunch. "The boys," as he invariably called them, now securely lodged in middle age—they ate out almost every day. An hour, two hours. Three or more, Ros calculated, if they ran into somebody interesting.

She learned to handle the intangible essentials that kept the small shop running. Distinguishing important calls from distractions. Retrieving Sam from the workbench when a customer's order hung in the balance. Unobtrusively heeding Sam's mood, often grave or agitated since the loss of Alfreda, and coaxing him toward a brighter prospect by indulging in small attentions such as filling the jar of Hersey's chocolates stationed on his desk, or plying him with a plate of fresh-cut figs from the garden. Knowing that it was figs he loved and wanted at that moment, and not pears.

Ros protected Sam, as Alfreda had done for decades.

"Sam had a very soft heart," remembered Ros. "He'd buy anything if he thought he was helping someone out." Which clearly wasn't in the interests of the business when it came to purchasing lumber that still hadn't been delivered or another round of superfluous items from any of the numerous salesmen who had identified Sam as an easy touch. When called upon, Ros could bring the harder edge to dealings that Sam refused to apply himself.

"We had a man who came in a van and sold rugs. Had done it for years. Pulled up, opened the back of the truck, and Sam would buy rugs. Finally one day, Sam said to me, 'You've got to handle this. I can't do it. I don't need any more rugs.'"

For decades, Alfreda had reminded visitors that Sam needed to return to work when the conversation lingered. Sam could then throw up his hands, as if to say, *What can I do? I'd love to sit here and talk with you all day, but you can see that she won't let me.* Sam counted on the luxury and the relief of having someone trusted and agreeably steely nearby to rescue him from himself. Raised by a sturdy, capable mother with his father frequently absent, surrounded at first by sisters and then throughout his life by women who found him charming and magnetic, Sam more than most men of his generation appreciated strong women with a light, directive, but insistent touch. Depended on them.

With time, Ros's admiration for Sam as an artist intensified as she perceived the subtleties of his work, the uncanny powers of his perception.

"He could walk in the house and see where a piece of pottery had been placed, and say, 'Oh, that's not right, somebody's moved it.'" Sam would alter its position. "And he was right. It did make a difference. He had this wonderful eye. As do the boys; the boys have that eye. To this day, they'll look at photographs of the furniture and say, 'This picture is better than that picture.' When to me, they're the same."

Ros also assumed guardianship of the secret history of the Maloof workshop—the comprehensive records particularizing the long, sometimes meandering, marvelously personal relationships with every individual and family that had ever ordered a piece of furniture. The archives contained file folders for each client, stuffed with forms, receipts, delivery notices, correspondence. Documentation of planning, pricing, responses to special needs and idiosyncratic stipulations. Memos tracking the foreseeably slow pace of fulfillment. But in many cases, the folders contained much more, far beyond what any other business would deem worthy and reasonable of assembling. Birth announcements, marriage invitations, funeral notices. Items culled from newspapers and magazines subsequent to completing orders. In bulk, they constituted a portrait of the artist in scrupulous engagement with his patrons, the quotidian detail entwining work and lives.

"Did Freda have the intuition that Sam was going to be famous?" wondered Ros. "Or was that just her nature? One of the joys of my job is that we're now getting a second generation who owns the furniture. Somebody comes with a table that needs repair, and I can say, 'Look at this file. This is your mother, or father, or grandmother's handwriting.' Sometimes they'll want copies. I don't give out the originals."

Surely, this is not how a business behaves. But Sam's clients often did not feel that they were negotiating with a business. Rather, they had entered into a friendship with the artist. The furniture remained merely the most palpable evidence of their accord.

Ironically, the business side of life suggested aspects of the man that his stature as an artist did not permit.

It's easy to imagine Sam chatting guilelessly with countless customers over the years, his spontaneity kindling expectations for an intimacy that was often robustly met. How surprising, then, to learn that whenever the shop sent out a new piece to a customer, Sam crept back into the business office within the week to anxiously ask Ros, "Did they call? Did they like it? Was it okay?" Even in his final years, an icon, he fretted. "Maybe you should call," he pled with Ros, "and find out."

"'Sam, if they're not happy,' I told him, 'we're going to hear from them. If they don't call, it's okay.' To me," concluded Ros, knowing the word she was about to say did not comport with the image or reputation, "it was insecurity."

Insecurity is something you won't find in the archives, though it perhaps remained an inscrutable component of the artist's process. Insecurity as an irritation, a prod. A reminder of his suspicion that there were improvements yet to be made.

"Up to the time he passed," recalled Ros, "he was working on three new designs. He was always creating. But every piece that went out, he'd ask me the same thing. 'Have we heard from them yet?' If they did call, I always had them talk to Sam."

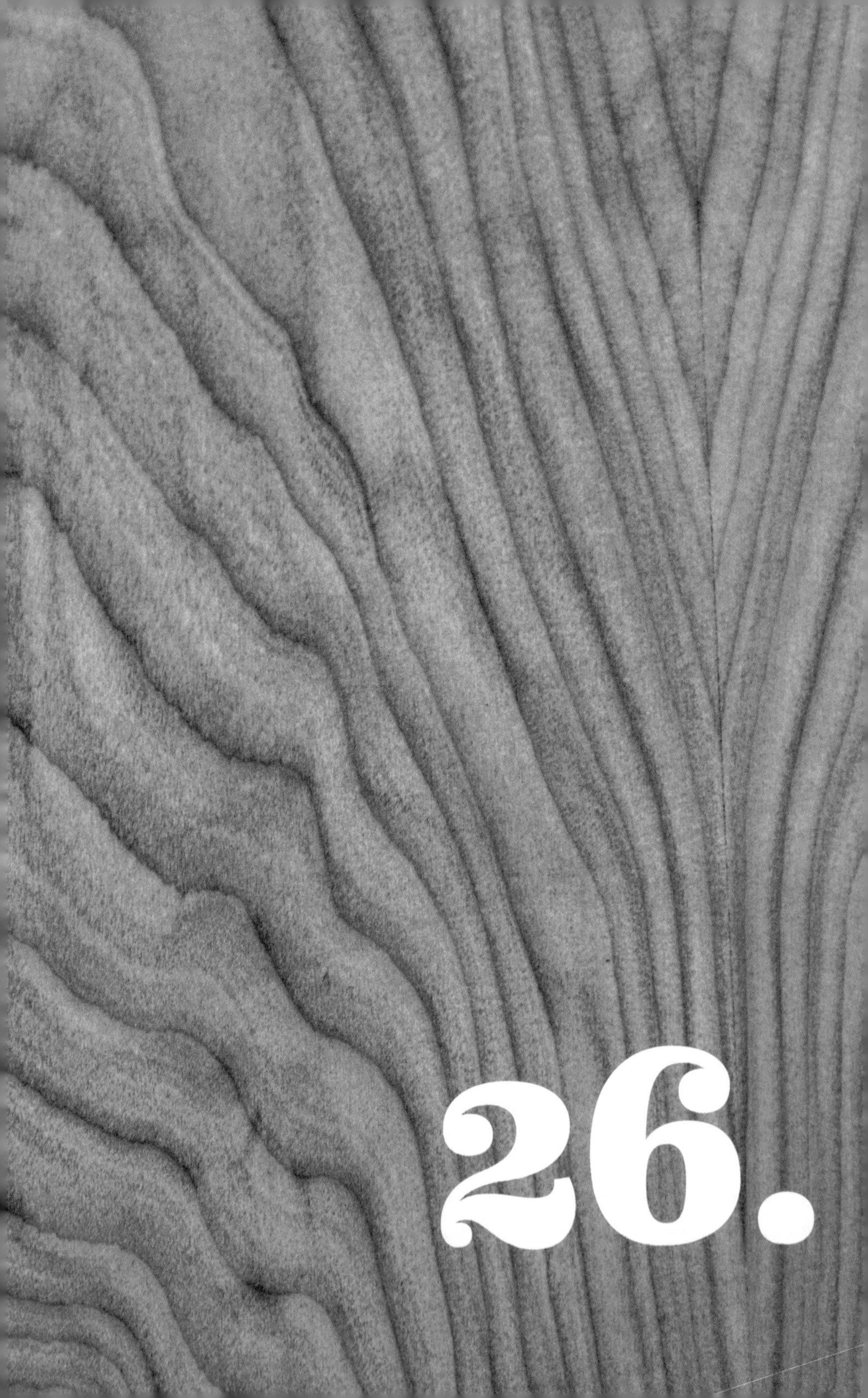

26.

PERFECTION
TED AND GLORIA MALOOF

Let's consider two recollections of Sam, the celebrated artist whose mere presence stirred notice and admiration:

He had arrived in Atlanta, visiting his friends Ted and Gloria Maloof, both photographers. (No relation—though it had been the shared last name that first caught Ted's eye in the mid-1960s while thumbing through a book about contemporary Californian artists and craftspeople. Shortly thereafter, a client in Ted's law practice asked if he knew Sam. "No, but I'd sure like to." The client picked up his phone and dialed Sam. When they met, Ted remembered, "It was like love at first sight.")

Piedmont Park was located a mile northeast of downtown Atlanta, the site of a huge annual art fair. Sam, Ted, and Gloria wound through the warren of booths where artists from throughout the region displayed their work. Somebody recognized the visitor.

"'Hey, that's Sam Maloof!' he shouted. They couldn't believe that Sam was there," remembered Ted.

A line formed behind him. Soon, it grew into a procession. "He was like a pied piper. All the woodworkers knew him, and a lot of woodworkers were at Piedmont Park."

There was another time in Southern California, Ted now visiting Sam. A tour of artists' homes in the Claremont area. The artists threw open their doors—occasionally to a modernist showplace designed by Richard Neutra or Fred McDowell, more often to a detached studio fixed in back. Sam and Ted joined the rounds, eventually landing at a woodworker's residence. The woodworker immediately recognized Sam.

"Sam Maloof! I can't believe it! You came to my house? Thank you, God. Thank you, God!"

"Something was always happening around Sam," acknowledged Ted. "I've had a much richer life having known him. Both our lives have been enhanced tremendously by being in his orbit."

"He made you feel that you were important," agreed Gloria. "He was kindness itself. But he expected to be looked up to and be the leader in whatever situation he was in. He dealt with prominent people within the craft world. But it didn't matter. He was always top guy."

And what was the source of his peers' esteem and the adoration of strangers? Ted and Gloria settled on a problematic word.

"He was obsessed with perfection," asserted Ted. "He had his own ideas about everything. He'd go out to his garden, and if the gardener had moved a little bush, Sam would notice it. Notice the bush had been moved. And he'd demand to know why, who authorized it? All about a little bush."

"It can sound like he was overbearing," said Gloria, "but he had a great generosity of spirit. Especially to those he thought were working really hard and trying to make something just right."

"I never have met anybody like him," said Ted. "He had a great sense of aesthetics. Everything he did was beautiful and meaningful."

But was it truly perfection that drew the woodworkers into the procession in Piedmont Park and thrilled the craftsman in Claremont? Did Sam imagine that he had achieved an ultimate level of craftsmanship?

"The low-back chair was the only design that reached its conclusion in his mind," observed one of his shop assistants. "The rockers he was still tweaking up to the end. His ideas kept evolving. He had the perseverance to pursue those ideas in an era when rapid change is the name of the game. To work with form, modify, keep making it better—that's what he was all about."

If there's a loftier goal than perfection, it must be the pursuit of perfection.

"He would not settle for anything less than the absolute best that he was capable of," said Ted, tightening his focus on the aspect of Sam's practice that earned the keenest admiration. His capacity to perceive minute flaws imperceptible to almost everybody else and then, after much effort, render them harmless. The willingness to never be satisfied with anything, while refusing to be mired in immobilizing dissatisfaction. It's the opposite of grandiosity; it's heroic compromise. The wedding of the practical with the ideal: the craftsman, with bills to pay and a worrisome deadline for the completion of yet another table or chair, remaining in bracing, troubling conversation with the artist, committed to spending the rest of his life courting the ultimate and unachievable. Those woodworkers in the park and those artists in their studios might well have cared less about what Sam achieved than the long, strenuous arc of his inexhaustible, bull-headed effort.

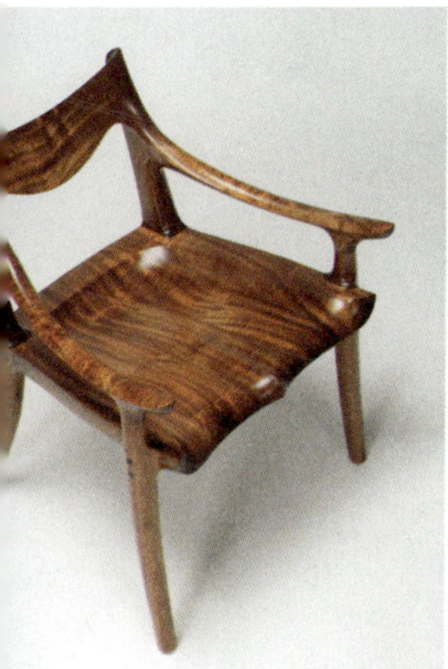

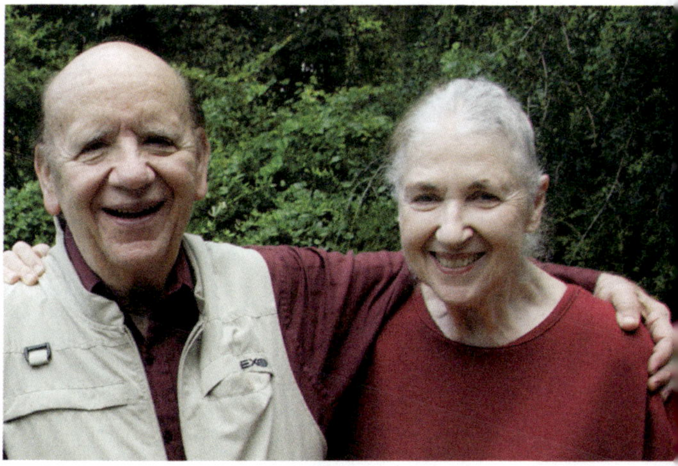

Left: A Maloof-designed ziricote chair. Right: Ted and Gloria.

27.

ADVICE
LARRY GANDSEY

At a certain age, indirection is a virtue.

Larry Gandsey completed two years of community college and figured he'd try not going to school for a while. In the early 1970s, picking up jobs proved a snap. For a year, he installed cable TV in homes throughout the East Bay, and when his boss nixed a request for two weeks' vacation instead of the customary ten days, he quit. He hauled his bike onto the Amtrak coach bound for Eugene, Oregon, and spent a month cruising down the coast and back home. In college he had played soccer, storming his way to the state championships with his teammates from Chile, Yemen, Portugal, and Russia; and so, naturally enough, Larry signed on part time as a coach at his old high school.

But surely there was something else to do with his life. Something larger in scope, more challenging, nourishing, perhaps even lucrative.

His roommate's brother ran the Butcher Block Shop, an East Bay furniture store opening just as the bulky, blocky East Coast style hit the West. The owner needed hands to build tables. Not skilled hands, necessarily, but hands belonging to somebody who would show up each morning and pitch into the work. Larry had never taken a single woodworking or shop class.

Hired.

"It was a very simple seventies design," he recalled. "Glue things together and apply a belt sander. Put a trestle base under it and call it a table."

During off hours, the crew could use the wood, workbench, and equipment to experiment with their own designs, building whatever struck their fancy.

Interesting, thought Larry. *You get to use your head and your hands at the same time.* He could picture himself doing this kind of work for a long, long while....

Three years later, the business shut down.

Larry had saved his money. He could take off a year and let the months unspool—perhaps travel across Europe, go who knows where? Something might come of it. He'd have an adventure.

"Or should I follow my heart in terms of what I really enjoyed right then and open a furniture maker's business?" That's what Larry asked himself. "The blessing and the curse was that I had never taken a business class."

He hadn't studied how to fail; he couldn't quite imagine how it might (indeed, probably would) happen. He was young, flexible, resilient. No encumbrances—no wife, family, house, mortgage, debts, or other intractable adult world worries. No need to restrict the scope of possibilities when they were aching to be enlarged. He signed up for some drafting and interior design classes, paid attention to what local craftsmen were building. He constructed furniture to clutter and smarten up his own home.

"That's what we do," he explained, speaking for the unofficial guild of young woodworkers in formation. "Because that's what we can afford."

In short, he "bumbled around"—his term. Meaning, in his case: growing more committed and working even harder as his skills developed. Steeping himself in tools and technique. Perking up his ears to prevailing fashions and customer preferences.

Striving to make a living, he made himself into a woodworker.

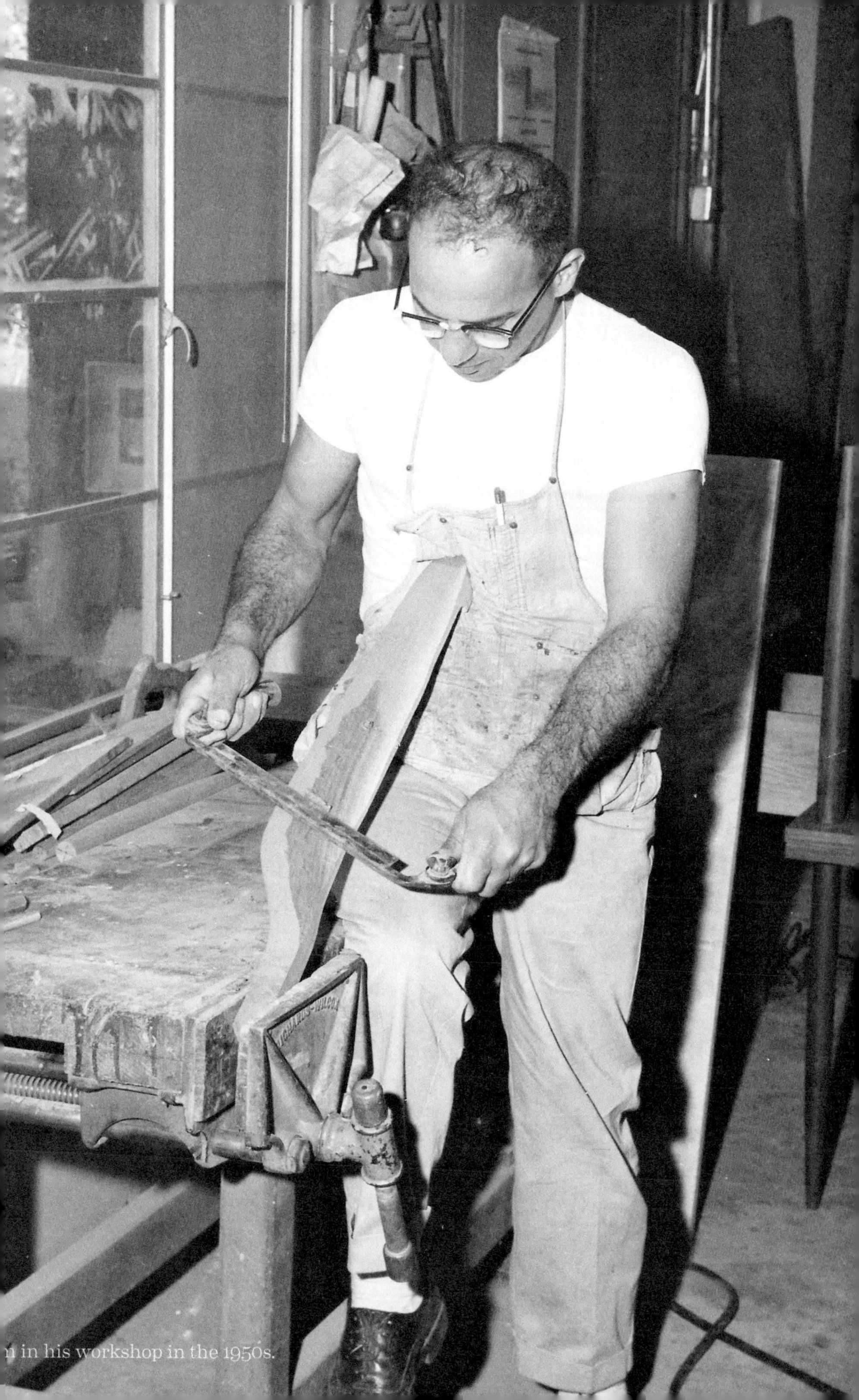

n in his workshop in the 1950s.

Today, Larry shares a four thousand–square-foot shop on Adeline Street in the industrial edge of Emeryville, California. He builds large, gorgeous conference tables and tables for restaurants, smart live-edge benches—mixes of maple, Monterey cypress, and sycamore; bay laurel, American elm, and eucalyptus; claro walnut, black acacia, cherry, and redwood. He's a full-time furniture maker who has managed to buy a house in overpriced Oakland, send two daughters to college, and continue his ongoing education in how to make useful objects for the workplace and home that cause the heart to ping with admiration.

How does an untutored young man full of unstructured yearning travel this distance to productive middle age? How does one slip into one's ambitions and carry them off gracefully?

"It starts with being aware of what other people are doing," he explained. "Which brings us to Sam."

Early in his career, Larry attended a symposium at the Oakland Museum entitled "Making It"—a nod to artisans in wood, fabric, clay, and other media who were managing to support themselves and their families.

Sam addressed the audience.

Of course, Larry knew him by reputation. Not yet a world figure—a decade would pass before his MacArthur fellowship—Sam nevertheless commanded attention, respect.

"He was featured a couple of times in *Fine Woodworking,"* said Larry. "Working for himself, making a living. He was *in the magazine!* That meant he was successful."

Here's what Sam said:

Follow what you're doing.

Certainly, Sam elaborated. (He always did.) The precise words have long faded. Yet the sense of Sam's advice rang clear and true in Larry's ears, then and now.

Follow what you're doing. There's a leading edge that you may not understand yet. (And there aren't any theories, much less a doctrine, that will explain where you're heading.) You know when it

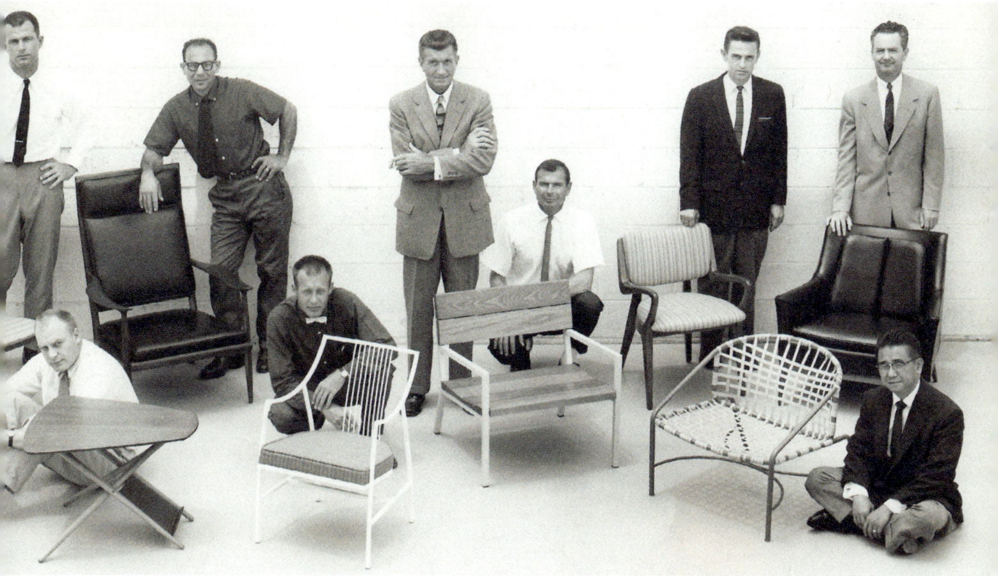

A 1958 *Los Angeles Times* article profiled Southern California furniture designers, including, from left to right: Kipp Stewart, Rex Goode, Sam Maloof, Stewart MacDougall, Hendrik Van Keppel, Taylor Green, John Keal, Barney Flagg, and Ernest Inouye.

feels good. Correct. You know. You have to live on the brink of your work, its promise, and your desire and ambition (however unstructured and inchoate they may feel at this moment—which is not the same as being formless and vague, as you certainly understand....) Follow what you're doing until ideas begin to emerge. They may change the course of your practice, nudge it a few degrees this direction or that. Proceed gradually, taking your own measure. Return to your hands, your heart, the center of yourself that you normally have no business talking about. Nobody else can lead the way. Follow yourself.

That's what Sam did. It's how he stumbled and grasped his way through the dimmest light, from the beginning of his earliest efforts until he began to prevail.

Larry understood that perfectly well over the months and years and then decades as he picked up his tools each morning and fashioned his own career.

28.

GIFTS
TONY ABEYTA

Picture Sam and Tony crisscrossing the LA Freeway, Sam at the wheel of his new Porsche Boxster. Sam's in his late seventies, early eighties—still fit and ceaselessly energetic. Tony Abeyta is thirty-five or thirty-six, Hollywood handsome, a painter from Santa Fe whom Sam has lately taken to calling "son."

Sam talks incessantly. Tony cares about everything Sam has to say. They discuss painting, architecture, wood, ceramics, art history. Sam veers off the freeway to show Tony a mural he worked on years past. They spool into Claremont to visit a favorite bookstore. In Los Angeles they have lunch with another artist or gallery owner. Over the years, during Tony's busy three- and four-day visits, they attend a procession of shows about American modernism, during which every time Sam steps onto the museum floor, the curators and even some of the visitors regard him as a celebrity.

"What do you think about Harry Bertoia?" asks Tony as they cruise down the highway.

"Oh, Harry Bertoia," nods Sam. They're heading back from San Diego where the young man and the older man had devoured a huge spread for lunch, visited a private collection, trolled through a museum—and still Sam volunteers to drive all the way back to Alta Loma so Tony can catch a needed hour's sleep. "Harry Bertoia. He did the *Sonambient.*"

"What's that?"

"The sculptures. You go like this with your hands." His hands lift briefly from the wheel to shimmer midair. "You're supposed to touch them. Most galleries won't let you."

And Tony reminds himself to look further into Harry Bertoia. Sometimes when Sam talks, he takes notes.

"Sam came magically into my life at the time I needed him most," Tony remembered. "I was in crisis for a long time. I'd lost my parents—my father and other favorite family members I lost to alcoholism. I'm an alcoholic myself, in recovery. Everything back then was very disrupted. I'd been doing what I'd been doing as painter for more than fifteen years and I was even having a ten-year retrospective at the Los Angeles County Museum of Art. I was making a living, didn't have to do any other work—but I was lost creatively. Didn't have a sense of why. Creating and selling, but no clarity about the direction. The minute Sam and I met, we connected. We had deep conversations immediately. 'You have to come back, you have to stay here.' And I thought, Yeah...."

Mutual friends had urged Tony to call on Sam while he was working on a show in LA. They informed him that Sam had known his parents as a young man in New Mexico, even collecting several of his father's paintings. They had met at the Santa Fe Indian School where Alfreda taught, and then maintained a lifelong conversation through an exchange of letters. Tony had never heard of Sam Maloof.

He withdrew the number from his wallet, picked up the phone, and called.

"Oh, Tony," responded Sam, "you've got to come out and visit. I have some of your father's paintings. Come out and we'll have lunch."

Whenever he recounts Sam's words, Tony's voice drops a half step to deploy a pitch-perfect imitation, his burbling tones filled with affection that might have been expressed from the core of either man.

"I'd spend time with Sam, and I'd always leave feeling I was special. All artists need heroes. Sam was mine. His direction was clear: always do your best, finish your projects, use the finest of sandpaper to perfect aspects of your work that most people wouldn't think really matter."

An indispensable message proffered at the right moment by an older artist who embodied what Tony Abeyta needed most at that time to deepen as a painter and a man. But perhaps the friendship, the trust suspended over a chasm of years, held a commensurate value for the person so generously offering it?

Sam had just lost Alfreda. He seemed overwhelmed by the plans to move to the new house and studio. His friends worried about him, perceived a creeping frailty, a ceding of curiosity and ardor. "These were the dark times," said Tony, "lasting two or three years. He seemed very, very lonely. Very depressed. He was broken then. But he worked through all of that."

Part of the passage involved a rekindling of memory, a steady recapitulation of accumulated blessings. And who better to listen than a young man whose appetite could not be sated?

"He talked about Alfreda every other minute," recalled Tony. "He'd tell the same story—I heard 260 times how he met her. Then I realized that when he was telling that story it wasn't because he thought I didn't know it. He liked reliving it. He had a visual way of thinking, his storytelling playing this little reel of memory. This vision of Alfreda standing there....'Oh, she was so lovely wearing a squash blossom and this beautiful dress. There she was, and I had to leave....' He told it again and again."

And while Sam frequently regarded the necessary disruptions of the move to his new quarters as an unwarranted wrenching away from the comforts of the past, Tony drew an altogether different lesson.

"I watched all that," he remembered, "and I thought: 'They're giving you this hillside with acreage and trees and relocating your house and building you a new studio and home?' I can't put this in terms for everyone to appreciate. But for the first time, I felt there was some order to things. They made sense. I felt divine orchestration. There was chaos, darkness, and birthed out of that came this really amazing thing"—the next steps in Sam's life.

Tony sensed the skepticism this proposition might engender. He paused, nodded to himself, and pressed on.

"I'm half Navajo," he explained. "We have this belief that if you're on a righteous, good path, then it's a pollen path. Corn pollen, one of the most sacred things for Navajo. You're in harmony with the universe. But there's also a dark path. More often than not, we're led astray there. Bushwhacking, you can get back where you need to go, but it will take longer. You're going to struggle, maybe with futility. You can get hurt. When you see things make sense, you follow that direction. The pollen path. I saw Sam's trajectory. I was able to bear witness to all those transitions, relocations, him getting married again. Everything so well orchestrated. And I thought: There's some director involved in this. God exists. That was one of the greater gifts Sam gave me."

Sam, the spiritual mentor, could also wax practical. When Tony was offered a $50,000 commission from a private collector for five large drawings, the prospect proved enticing, almost irresistible. But from the start, warning signs arose. When Tony pulled up to the collector's house, he had to navigate around a fan of black Mercedes convertibles and sedans—notification he was entering unfamiliar, possibly incompatible territory. Then negotiations began to break down. The family wanted the dimensions of his pieces to expand while the price remained constant. They pressured him to hurry along the drawings, keen for their unveiling at a party. Then the collector cut the fee paid to the decorator, who in turn wanted to pare back Tony's commission.

"Sam said: 'Oh, son, no. You can't do that. You'll run into more problems with them. People who want to take but don't respect the work. They don't respect you as an artist. They're just trying to fill wall space.' So I didn't do the commission. I didn't want to disappoint Sam. He was really sure it wasn't a good idea. I had to trust somebody. I trusted his opinion."

Sometimes Sam weighed in with personal advice, unabashedly answering questions that Tony was not quite posing.

"After my divorce, I'd bring different girls to his house all the time. One month, a Russian girlfriend. Another month, the Japanese girlfriend or the hippie with curly hair." Tony recalls his most audacious introduction: "'Sam,' I said, 'I want you to meet the love of my life.'"

Sam eyed her sympathetically, but realistically.

"Oh, no, no, son. She's not."

Yet the most intimate conversations always circled back to art and the world of artists in which Sam had thrived.

"I'd ask, 'Sam, what do you think about Beatrice Wood? Should I get a Beatrice Wood?' Sam taught me everything I needed to know about studio ceramics, which I then started collecting as a result. 'Oh, Tony, I don't know....Let me say this. She was refined. She invented this lusterware glaze. I knew her and...'—and he'd tell me these great stories."

Through Sam, Tony met Harrison McIntosh, Rupert Deese. "Sam talked about him and Millard Sheets printing silk screens and Millard working with the muralist, Ramos Martinez. Sam's eye was very informed. He devoured everything. He'd come out to Santa Fe and call. 'Tony, you're having breakfast with me and Bob Turner.' I'd seen some of Bob's ceramics and now I got to meet him. I was the ideal friend to have. Everything Sam said, I'd listen and ask him questions. I was interested in that period of California, how artists were creating, how their careers went in different directions. I'd question him about this and that, and he'd give me the most personal answers."

When Sam was dying, Tony was holed up in Santa Fe, working fiercely on some paintings. One evening he called Beverly, and when he learned that Sam's condition appeared critical, he volunteered to fly out in the morning. But Sam wouldn't last until morning. Bev fit the telephone up to Sam's shoulder, nestled the receiver at his ear.

"'I want to thank you for this, and for this, and for this,' I told him. Ten minutes, and I'm still not done yet. 'Thank you, Sam, for this, and this, and...' Fifteen minutes. I was manic. I kept thanking him. I did it for me. 'Thank you, Sam, for the spiritual gifts that came out of all this....'"

We like to imagine that the contributions we make during our lifetime retain some aspect of their worth after our passing. That in the most fortunate and special cases, they might even multiply.

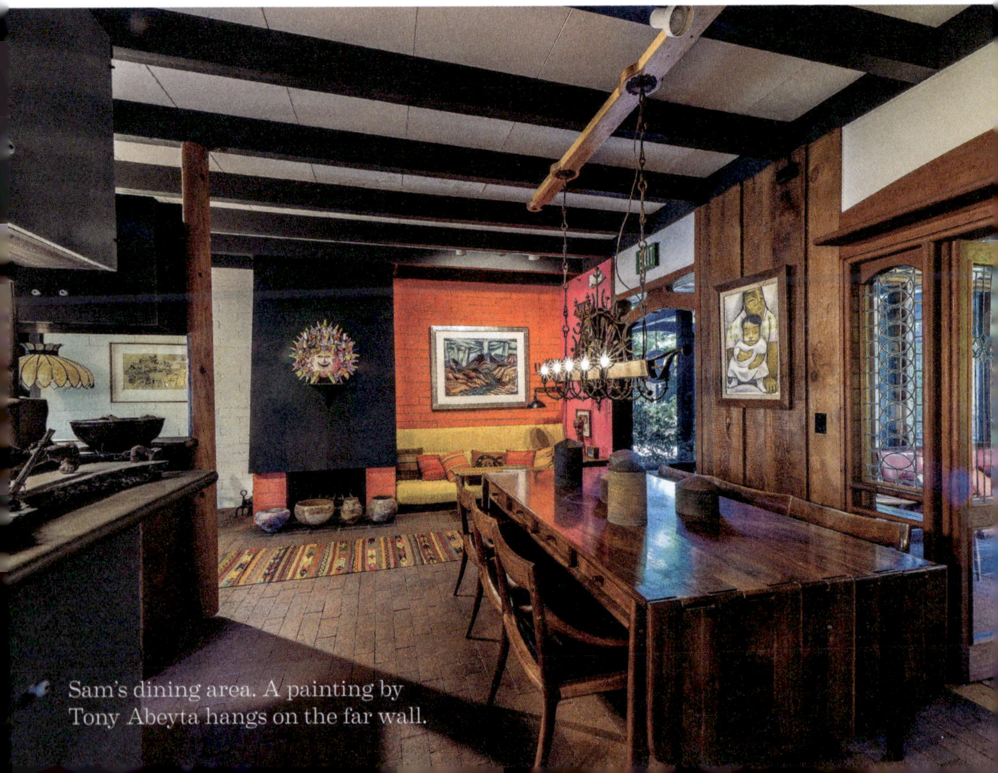

Sam's dining area. A painting by Tony Abeyta hangs on the far wall.

In the saga of Sam and Tony's friendship, this sanguine prospect seems evident.

Today Tony guides and advises students of his own, expanding on the promise and snares of an artist's career. Perhaps he even points the way toward a pollen path. And whenever a stranger crosses the threshold of his studio and begins a conversation, he embraces the possibilities, grips them. *C'mon, sit down. Let's talk. Where are you from? Why are you here, what do you see, what's driving you, what do you yearn for, where are you going?*

"I try to echo Sam's ideas about contribution. 'Tony,' he told me, 'you have to give back. You have to give in order to receive.' That's something I really heard and it stuck with me. Without becoming Sam—I'm my own unique person—there are some of the doctrines I follow."

Step into Tony's studio these days and you might see him working on one of his huge canvasses featuring a profusion of flowers. Aerial depictions of a pocked bombscape with the blossoms bursting into animation. Some of the canvasses are awash in detonated yellows, oranges, and reds; the majority in an equivalently discharged black and white. Each painting a decisive parting from his earlier work.

Moloch Blossoms, he calls them. In respectful reference to Allen Ginsberg's *Howl,* a kind of holy text.

"This is what masculine energy feels like," he said. "Volatility. I want that explosive nature of the raw, beautiful male."

Or, alternately, he deems the paintings Testosterone Flowers.

The father, the son. The son and father. The man.

"I always said this about Sam: He works with the hands of God. Pure divine channeled through. It was no accident we became friends. But it wasn't just me. I'm not remarkably special in this story of Sam. I'm one of, say, sixty people he's had a direct effect on, in really positive, creative ways. Where did he get that generosity? I just think he knew that he had a lot to give."

29.

SAM'S WOOD

For a couple of years after the move, Sam kept his lumber stores under a makeshift blanket of tarps positioned on the small patch of ground now occupied by the gallery. Finally, he built four storage sheds, doling out his own money to complete an addition to the property whose necessity became increasingly evident every time a flatbed truck rolled through the gate and began off-loading another contribution to the collection.

High-piled lifts of cherry and maple. Oak. Tasmanian rose myrtle.

Sam liked to tell friends that he collected wood as "a passionate hobby." He also considered the rising stacks of lumber a shrewd investment, like prime real estate—"better than money in the bank."

Mike Johnson and Larry White, longtime coworkers in the shop, undertook the organization. Sam would often dispatch them to select or at least identify lumber for an upcoming project. Years of sifting through piles to locate an ideal slab at the bottom of the heap persuaded them to construct bins for vertical storage, separating the wood by both species and dimensions. Today, visitors negotiate a path through the sheds with the feeling of entering a dense, towering forest of the sawed and planed.

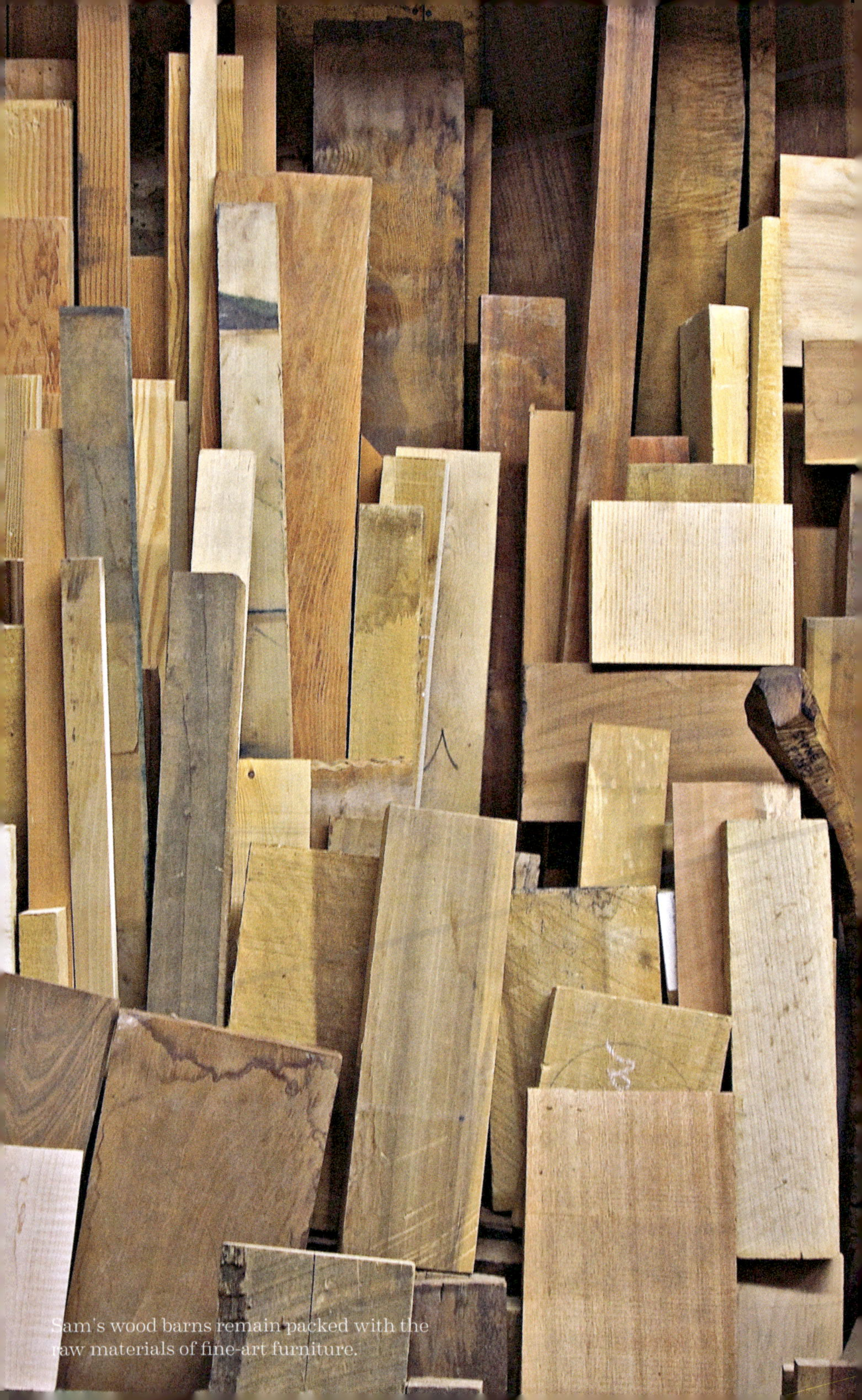

Sam's wood barns remain packed with the raw materials of fine-art furniture.

Mahogany, African ebony, Macassar ebony. Claro walnut, eastern black walnut. Ziricote.

In the corners and far walls of the sheds stand an array of woodworking machines. Band saws. Lathes. Drill presses. Planers, routers, joiners. Essential tools cast off in multiples. Sometimes Sam received machines in trade for lumber. Some he acquired and never used. Sam never threw away anything. Everything had sentimental value; every piece had its own story. At the old site, he had surrounded several trees with piles of machinery that struck the uninformed eye as junk; he deemed them essential. Sam had grown up during the Great Depression; he was a citizen of the nondisposable world.

Peek into the bins stuffed with figured maple. A distinctive grain pattern identifies each variety. Quilted maple simulates rippling water. Stripes mark the species referred to variously as curly, flame, tiger, or fiddleback maple—a hardwood frequently employed for the underside of violin bodies. Much of the lot has been graded by its vendors. (Curly Maple, Quadruple A, prized by the manufacturers of hard-body electric guitars.) Whenever Mike stumbles upon a piece that suggests a future use, he adds a few chalk marks to later jog his memory. Amid the incalculable tons of lumber, hung high above the bins, a reminder of a different kind: a poster for Sam's 2001 show at the Smithsonian, *King of the Rockers.* At the center of the poster, a rocking chair crafted from fiddleback maple draws the eye.

Wood has to dry before being used for furniture. Otherwise, the joinery can come apart as the fibrous tissues lose moisture and shrink. A chair or table might end up over months or years disassembling itself. Patience is required: wood dries at a rate of one inch of thickness per year. A two-inch-thick piece of lumber destined to serve as a chair back might take two years to dry. Some wood is kiln dried at the mill, the moisture content lowered to a safe rate for immediate use upon delivery. Today, all the wood in the Maloof storage bins is dry and ready for work.

Sam deemed black walnut his favorite wood. He prized its "workability"—easy enough to shape, yet sufficiently hard to maintain the crisp, definite lines of his later designs. Perhaps 75 percent of his furniture employed black walnut. The bins still contain abundant stores of walnut, including the gnarly, twisted grain of one immense slab of native California claro walnut cut close to the base. Growing in the Chico valley north of Sacramento, the species is a valuable hardwood sought not only by furniture makers, but also the shapers of gunstocks and fine auto veneers.

The sheds also contain plentiful exotics. Ziricote from Belize. Ebony from Africa and Malaysia. Huge pieces of Honduran mahogany. Edging a pathway through the exotics is like visiting an art gallery filled with warm, expressive abstractions. African Bubinga with its three-dimensional waterfall effect. Brazilian and Indian rosewood. Some exotics are now rare, impossible to import because

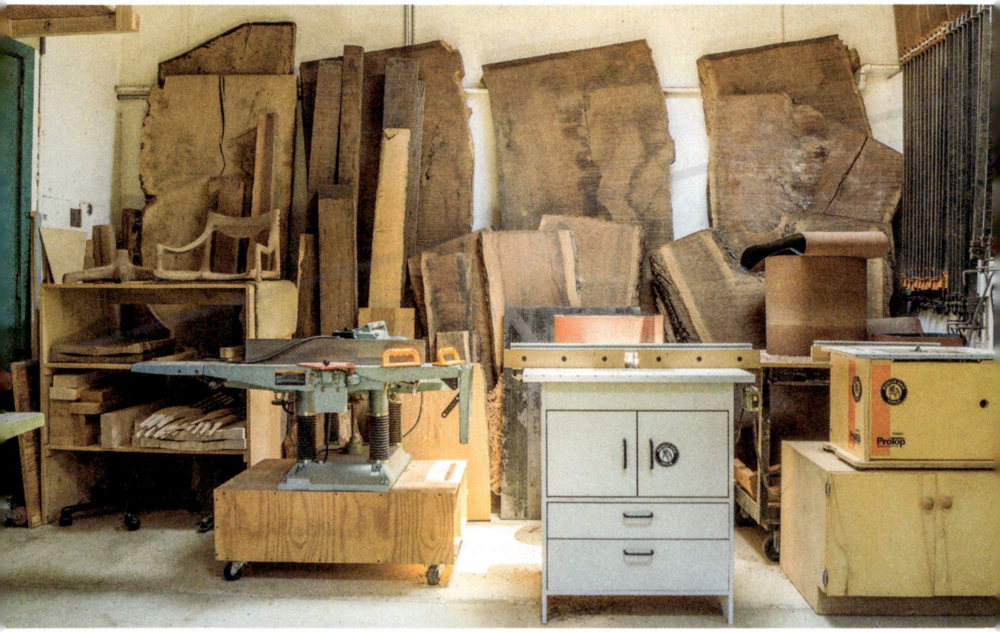

of embargoes on raw materials. One irreplaceable piece: the wood cut from an enormous avocado tree that stood on Sam's original property. Sam loved the tree. It grew to six feet in diameter, the giant of his numerous avocado varieties. He moved the toppled log to the new site where it was milled, stored, and now (like so many other pieces) awaits service in the ideal project.

Many bins contain valuable scraps—priceless resources for a craftsperson making small items. While Sam was still living, Mike Johnson undertook a series of miniatures of shop furniture, designing each table, chair, and rocker to quarter scale and fashioning the joinery to precisely reflect the look of the originals. When photographed, the miniatures had to be positioned next to a thimble or coin so they could be distinguished from their larger kin.

Sam purchased most of his domestic wood (walnut, maple, cherry, oak) from the Peterman Lumber Company in Fontana, a dozen miles from the shop. Sam and Pete Peterman were friends for years. Pete might call with the news about a mill located somewhere in the Midwest about to dispatch several prime lifts of beautiful wood by train, and Sam would buy a portion of the load sight unseen. Toward the end of his life, Sam purchased increasing amounts of lumber from varied sources, collecting rare, beautiful varieties with the same enthusiasm that he applied in acquiring ceramics and paintings. Today, more wood is housed on the grounds than could ever be used in the productive lives of the shop's craftsmen.

And still the offers keep coming.

Just the other day, the shop fielded a call from Canada. A miller had a new supply of big leaf western maple, another fine, highly figured, light reddish-brown grain particularly well suited for furniture. Just a few years ago, when Sam was still alive, the miller's old flatbed truck, loaded down with twice as much as it should haul, sometimes showed up at the gate unannounced, the run made all the way from Canada in hopes that Sam would buy some.

He'd often buy a lot.

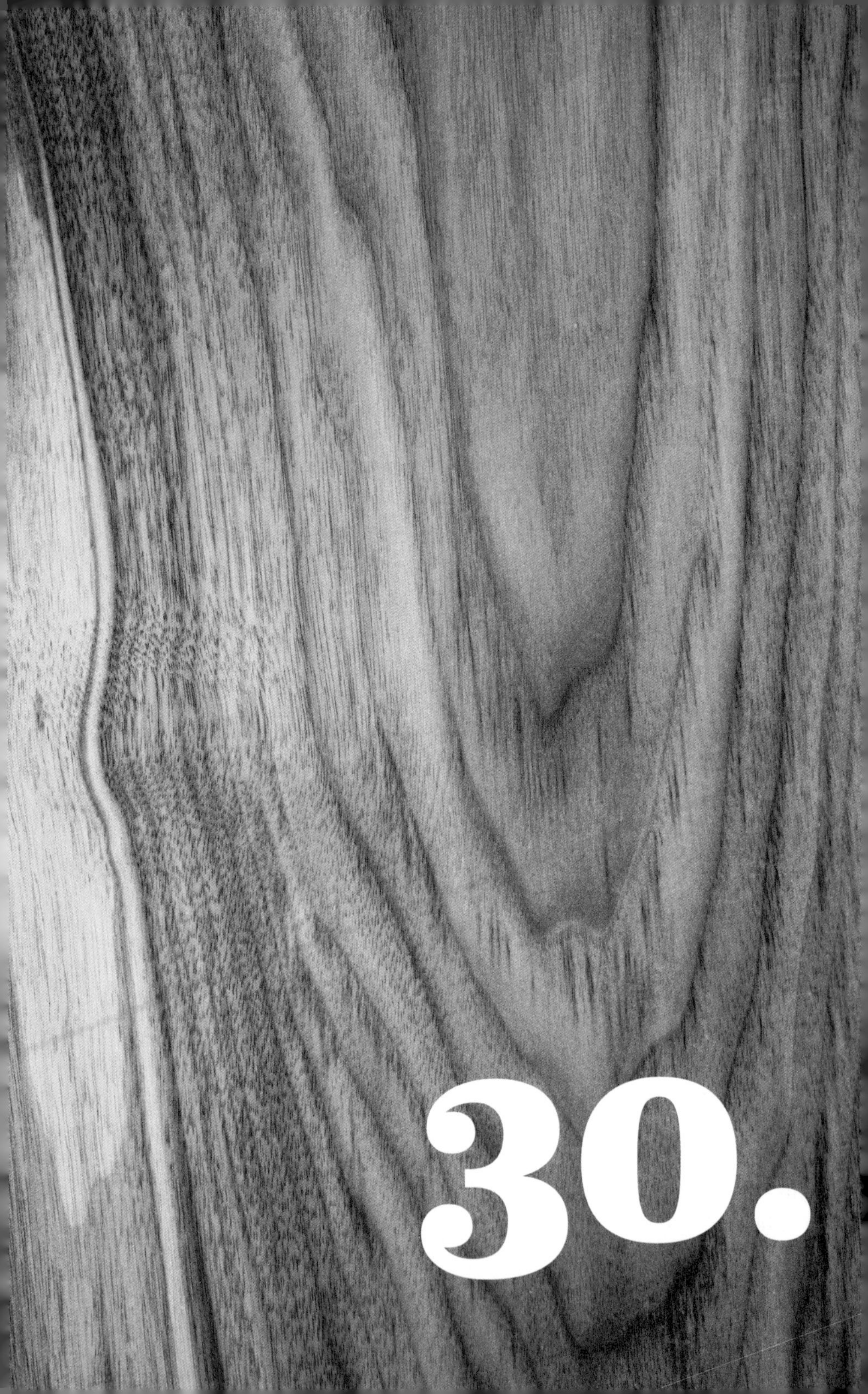

30.

MEMORY
CAROL SAUVION

The camera strokes the rocker's curve, following the arm as it slides into the spindled back, rendering the form of the chair surprisingly full and substantial for its two-dimensional pixilated assembly on the screen.

"I want a chair to be very beautiful," says Sam, his voice weaving into the camera's patient long shots. He actually says "bee-yoo-tee-ful"—Sam's unique pronunciation of the word typically applied by the rest of the world to describe his furniture. "I want a chair that's good to the eye. I want a chair above all that's comfortable."

It's the initial episode of *Craft in America,* the fourteen-part PBS documentary series on contemporary crafts-people and their diverse approaches to the creative use of wood, clay, cloth, and a host of other materials, produced by Carol Sauvion and first aired in 2007.

"I've had people say that my furniture is art," Sam continues as the camera's gaze comes to rest on the seat of his rocker. "That it's sculpture, it's this and that. But I really don't care what they call it as long as I'm happy with what it is."

The filmmakers have titled their episode "Memory," but Sam is still very much with us. The camera trails behind as he bustles into the shop, somewhat slower and more deliberate in his tenth decade. At a worktable, his assistant sands a fiddleback maple rocker. "It's going to be a knockout," acknowledges Sam, pressing on through the shop to inspect a set of dining room chairs.

"I'm getting too old for this," he states without a mote of conviction, and then sweeps up an armful of wood and totes it across the shop floor. "I should lie down once in a while, but I don't." He chortles—a word that should almost never be used to describe human laughter; but he truly does chortle. "I love what I do. I'm ninety and I can still work." He spits out a laugh and applies his sander to the wood. "We do everything to order," Sam explains to the camera, his portal to the PBS audience, the millions of viewers who will never set foot in his Alta Loma workshop. "It isn't a production shop. We don't have parts all over the place that we're waiting to put together."

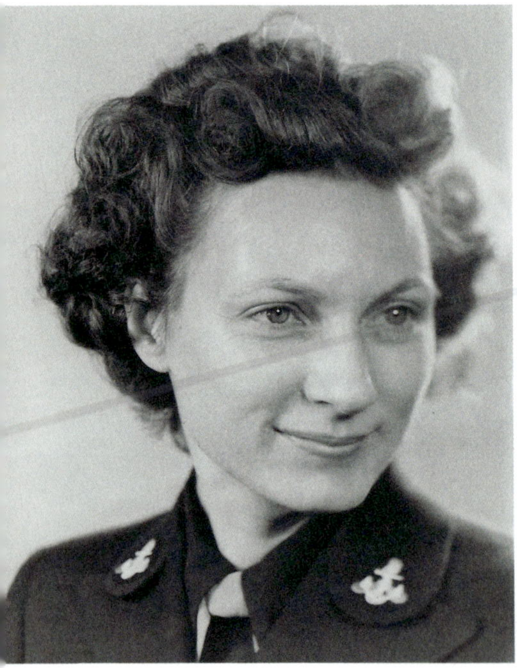

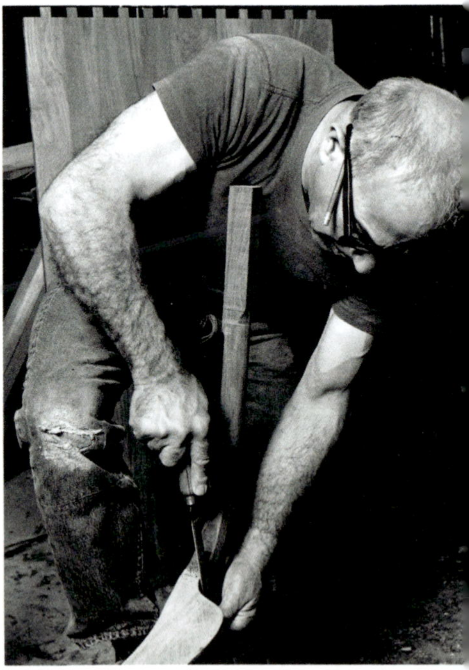

Alfreda Ward as US Navy Wave during World War II, before she met Sam.

Precisely the point. Producer Carol Sauvion began work on *Craft in America* in 1996 to alert the world to the immense and varied art of "the handmade." She acknowledges that the craft world "is marginalized and probably always has been. It's an economic issue. To make something by hand is more costly." Which has never mattered much to "someone like Sam who's self-taught. Who didn't go to university. He didn't even apprentice with anyone. He just figured it out. He's like a gifted genius"—propelled from the start by a readiness to court risk and failure, which never intimidated him as it might have had he apprenticed or attended the university. "Failure is essential to a craftsperson," said Carol. "That's a way to learn. You can't get it right all the time. But every time you work seriously, you learn something. Certainly, that was true of Sam."

We were speaking over the phone, our voices disembodied and hundreds of miles apart, while Sam continued to talk and gesture from the counterfeit distance of years past.

"When you're working," Sam explains, a shade of himself on the television screen, "there's a communion between the object maker and the materials he or she is working with. Then it transcends into something much greater...and when someone likes it and enjoys it and all, you're paid tenfold."

The camera envelops Sam in a characteristic pose. He's sitting, eyes set slightly above the camera's gape; he wears a brown leather jacket, black T-shirt, jeans. Occasionally, his big left hand settles upon his cheek in a gesture reminiscent of Jack Benny, the radio and TV comedian from Sam's youth.

"Scripps Graduate School," he recalls dreamily. "I was at the school one day, in the courtyard. I saw this girl. I looked at her. I thought, That's the most beautiful girl I have ever, ever seen. She walked all across the courtyard and said, 'Pardon, but can you tell me where the office is?' I said, 'I'm not going to tell you, but I'll take you.' I'd go back about twice a week and never saw her. Come September, I was at the same spot; she entered at the same spot.

Walked over to me and said, 'Hello, Sammy.' Three months later we were married."

Sam's story about first meeting Alfreda: toward the end of his life, he relived the encounter endlessly for both strangers who would be flattered by the implied intimacy and friends who could repeat the tale word-for-word. Sam spoke with an urgent, melancholy hopefulness, as though he could extract from the bundled fibers of his memory the corporeal witness of his beloved, absent wife. Recollection allowed him to visit the past daily, drawing the core of his life's meaning into the present.

"Freda was the heart and soul of what I do," he tells the American viewing public. "She became my partner in more ways than one. She was sort of my protector. If it hadn't been for her, I think I would have quit a long time ago. She said, 'We can do it, we can do it.' So we did it together."

"They had a purpose that was bigger than both of them," agreed Carol. "It was nothing they could put their finger on"—there were no Maloof proclamations about the purpose of art or the responsibility of artists. "They were just going to do it."

The outcome of their unarticulated ambition is evoked with every turn of the camera as we wander through the house. The camera glances over the paintings traded and purchased from Sam and Alfreda's friends, the pots and sculptures they collected together over the decades. It twirls around the double music stand Sam made for Jan Hlinka, the Los Angeles Philharmonic's first violist; the cradle swinging from an absent someone's insinuating prod; the rocker positioned at the center of a Navajo rug; the stained-glass door hitched shut by Sam's glossy latch.

"The goal," explained Carol, "isn't about getting rich or famous. The goal—and I find this is true for many artists—is being able to do what they want with their lives. Ideals and integrity—there's something about working with your hands that makes those things part of your life. It's not something you've got; it's something you live."

The camera peers out the upstairs window of Sam's house. "I followed what I wanted to do," Sam affirms. "I didn't let what was popular guide me." The screen radiates a glimpse of the grounds and garden before the image dissolves and Sam's world disappears.

"Even if you become a Sam Maloof," said Carol, "you still have to remain passionate. Many artists retire, or they become teachers, or find a different use for their knowledge and skill. Sam was creating up to the last minute he was alive. He was heavily reliant in the final years on the men who worked with him, but he was still involved. That, to me, is a true artist: You have to make things. You don't have a choice."

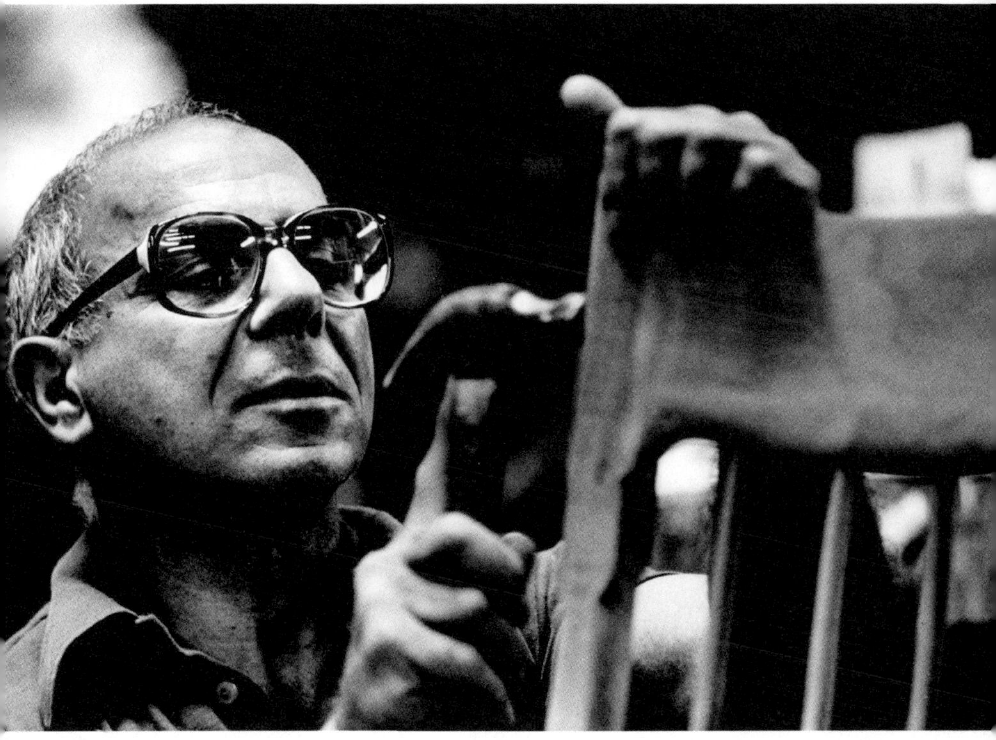

31.

TOMORROW
PRISCILLA SIFUENTES

"Sometimes, art doesn't have a why."

This seems an astute and rather subtle aesthetic judgment to be rendered by a twenty-year-old high school graduate. But over the past three years, starting in twelfth grade, Priscilla Sifuentes has devoted scores of hours to looking at and learning about the ceramics and paintings that fill the Maloof residence. Inspecting the sculpture that trails across its grounds. Studying, touching, admiring, and wondering about the furniture emerging from the shop.

As a member of the Maloof Teens, Priscilla is charged with introducing local high school students to their region's most celebrated artist. She visits their classrooms to deliver an energetic summary of Sam's career, suggesting that a visit to the shop and residence may well provide a deeper understanding about how art gets made, who artists really are, and what the appreciation and creation of art can mean to any of us.

Many of the students who show up at the gate come from immigrant families, like Sam's, where a language other than English is spoken at home. Most of the students' parents, like Sam, never attended college. None of the young people participating in the program have ever imagined, much less been welcomed into, a house quite like the one nestled into the hills of Alta Loma.

"When I first came here," said Priscilla, "I was struck by the calmness of everything. The serene exterior. But on the interior—Oh, my gosh! What is all this?"

Priscilla felt what the students she now leads through the grounds report feeling as they encounter the potent energy of the shop. The head-spinning variety of pots, paintings, wood, and metal sculpture—a kind of candy shop for the hungry eye. The peculiar allure of the wonderful house, not much larger in its footprint than the modest bungalows in blue-collar Ontario and Fontana, but evidently crafted by an artist's hand wielding the sharpened edge of soulful ambition.

This is what some students begin to realize, if only for an instant: Life can be lived as a beautiful adventure. But you must lift

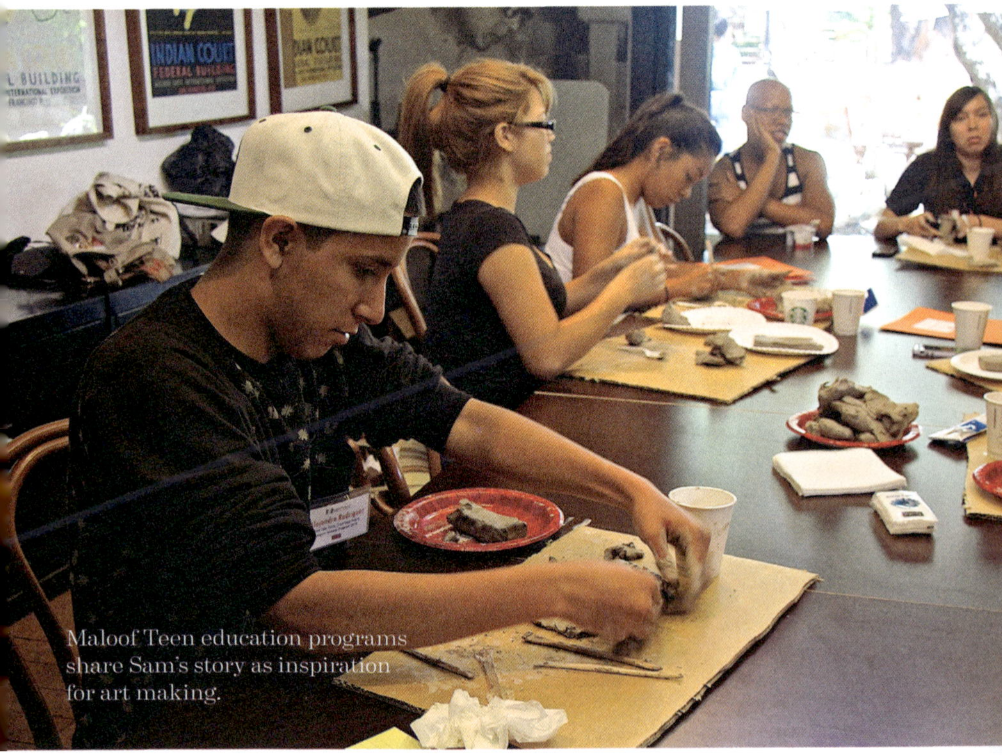

Maloof Teen education programs share Sam's story as inspiration for art making.

your eyes above the everyday. You must loosen your imagination, shoulder your way forward, and learn to work like a calm and happy demon.

When Priscilla first enlisted as a Maloof Teen, she thought of herself as "super shy." She wanted to change, to shake off the adolescent shackles of fear and self-consciousness. Uncharacteristically, she stepped forward and volunteered to lead tours after mastering a set of basic public-speaking skills.

"I can speak in front of classrooms now. Kids that I've known since elementary school, they'd hear me and be—'You talk? That's your voice?' If I give a general tour these days to the public, I'm talking to adults. Twenty-year-olds, thirty-year-olds. Seventy-year-olds! I never thought I'd be that comfortable." Earlier this year, Priscilla and several other teen advisors made a presentation to a corps of older docents so they could outline their own work in the schools. Priscilla's sophomore English teacher attended. "I was surprised she was here and she was surprised I was here—and talking. Later, she mentioned how in school I would try to negotiate with her to make my project presentation times shorter because I didn't want to be talking in front of the class. Now she was so proud of me."

Whether addressing high school students, her peers, or adults, Priscilla knew that she was making a momentous statement: I exist. I have something to say.

Just like Sam.

"Yes, he loved to tell stories," said Priscilla, who has heard countless accounts of garrulous Sam. "I love that I'm able to tell his story."

Telling Sam's story these days, Priscilla also tells her own. How "so shy" a young woman slipped unexpectedly into a new world and discovered her aptitude for curiosity, discovery. "Every time I come here, I notice something I've never noticed before." The Oaxacan wood carvings that remind her of her family's home.

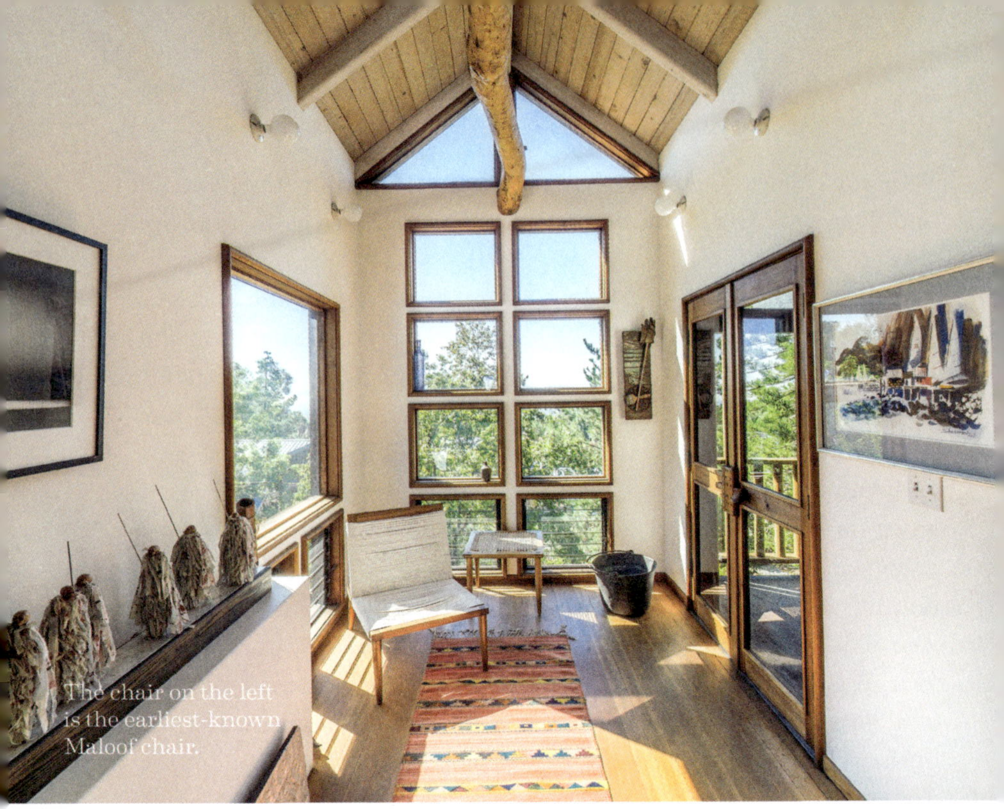

The chair on the left is the earliest-known Maloof chair.

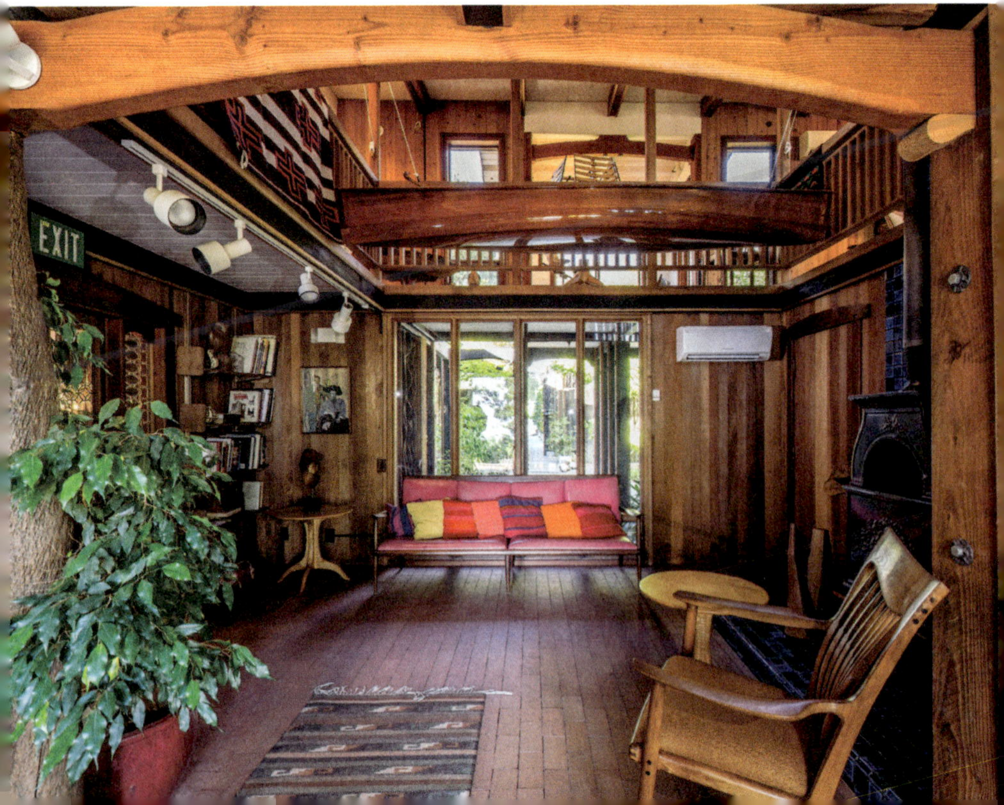

Or Millard Sheets's watercolor depicting a pair of Mexican women toting calla lilies, baskets of fish, and a baby. "I want to learn more about it. I want to find out what else is here. I've that hunger for more. More knowledge."

Soon, Priscilla will enroll in college. Perhaps become a teacher (as so many adults keep suggesting). Or maybe a photographer. (She's taken some classes, returns to her camera as often as she can between juggling three part-time jobs and volunteering at the Maloof compound.) Sometimes when she contemplates the gentle, rising curve of a Maloof chair arm, she'll think about Edward Weston's photos, particularly the rounded torsos of peppers, and how she'd like to capture some of that same shape of feeling. Perhaps she'll enter a museum studies program, a field she'd never heard of before joining the Maloof Teens. Become a curator. Maybe run a museum someday. The world has become an endlessly interesting place. Who knows what she might accomplish?

Sometimes art does have a why. With luck and timing, art can point us in the direction of a daring life. It can persuade us to settle for nothing less than the beautiful.

Sometimes, admits Priscilla, who first visited Sam's house two years after he had died, she thinks about him as the uncle she never knew.

"It's like he's right around the corner," she says. "I haven't met him, but I'm going to see him soon. He created all this and he's still here. You feel him right here."

32.

EULOGY

From Mike Johnson's eulogy at Sam's memorial service, June 9, 2009, in the Kresge Chapel of the Claremont School of Technology in Claremont, California.

"Let me show you boys what I am doing" were familiar words that often rang out from the room that Sam worked in. So with my coworkers Larry White and David Wade, we would come in and have Sam show us a new idea or shape that he was considering. Sometimes he was confident and sure; other times, he was open to the exchange of ideas and might ask for our opinions; and sometimes, he even used our suggestions.

Sam had a "learn by doing" philosophy. Usually with just a few words of instruction, he would then let us go to see how we would do on our own. I remember Sam telling me, "Keep one eye on your work and one eye on me; that way you will learn what I am doing, too." Well, I soon realized that Sam was keeping an eye on me, also. It always amazed me how, from across the room, he could say, "Mike, I think a little more needs to be taken off"—and after walking over to the piece that I was shaping, he would finish his sentence—"right here," pointing to the spot that needed more work. Then, I would stand back, look, and realize he was right.

Sam had an incredible eye for shape and form. In some ways, this may be his greatest gift to the world. For him, creating these forms came easy and natural. For me, mastering what he wanted took time. So since we were learning by doing, there was more shaping, and more sanding, and more constructive scrutiny. And more shaping, and more sanding, and more time....Then, gradually, at some point this changed to passing nods of approval and words like, "You're doing a beautiful job."

Many years ago Sam was breaking David in on grinding out chair seats. David made the comment, "What if I ruin one?" Sam replied, "Well, you might ruin several, but you'll never learn until

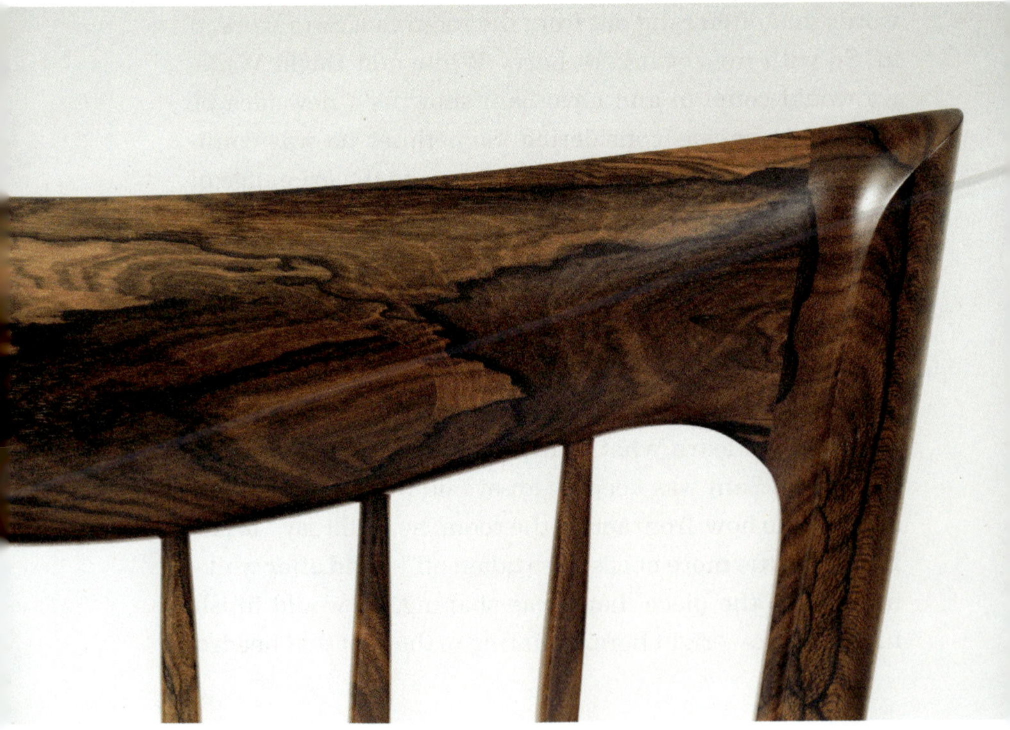

you try." Sam was patient with us and made it clear that he would rather have the work done right than fast.

With the sensuous curves of Sam's designs, the band saw was unquestionably the most important tool in his shop. When he was using it, you did not disturb him. Phone calls would need to wait. Sam was in his element. This was "his" machine, and he "owned it." I'm speaking about ownership in the same way that an athlete, who has risen to the top of his sport because of his skill and ability, "owns" the field he plays on.

The way that he used the machine, in what was a sort of "free-hand cutting" style, or some may call "band saw sculpting," has become legendary among the woodworking world. It brought many people to his woodworking classes, just to see him in action. The "oohs and ahs" from his amazed audience were priceless to those of us who were accustomed to his technique. You see, Sam was self-taught, and some of his methods were quite dangerous, especially for a novice. So, he wisely began by telling his classes "not to try this at home."

Yes, Sam was a showman, and he loved to perform. Often, when walking guests through the shop, he would stop at the band saw, grab up a piece of wood with everyone crowded around, and cut out a chair arm. In not more than two minutes, the whole thing could be completely roughed out using only the band saw. Then, he would hand it to a lucky guest as a souvenir.

33.

EULOGY II

From "A Visionary Woodworker" by Ralph Nader.

The exquisite manual workmanship of Mr. Maloof is further stimulating the questioning of the remoteness that modern technology visits on so many people who spend hours in virtual reality, separated from nature and its materials. Our country was built by craftsmen, artisans and other workers who designed and made real things. High schools offered Shop Class, where students learned skills and the joy of creating. These classes opened doors to a source of livelihood and pride for budding artisans....

In a new book titled *Shop Class as Soulcraft* by Matthew B. Crawford, the author bemoaned the closing at high schools all over the United States of shop classes that taught the mechanical arts like carpentry, woodworking, welding and other skills. They were closed to allow more funding of computer labs. And because our throw-away society no longer properly values the fruits of artisan labor. Crawford goes on to argue and demonstrate what our society loses when we make joining the paper economy the chief aspiration of the younger generations, or to use Robert Reich's phrase, to become "symbolic analysts."

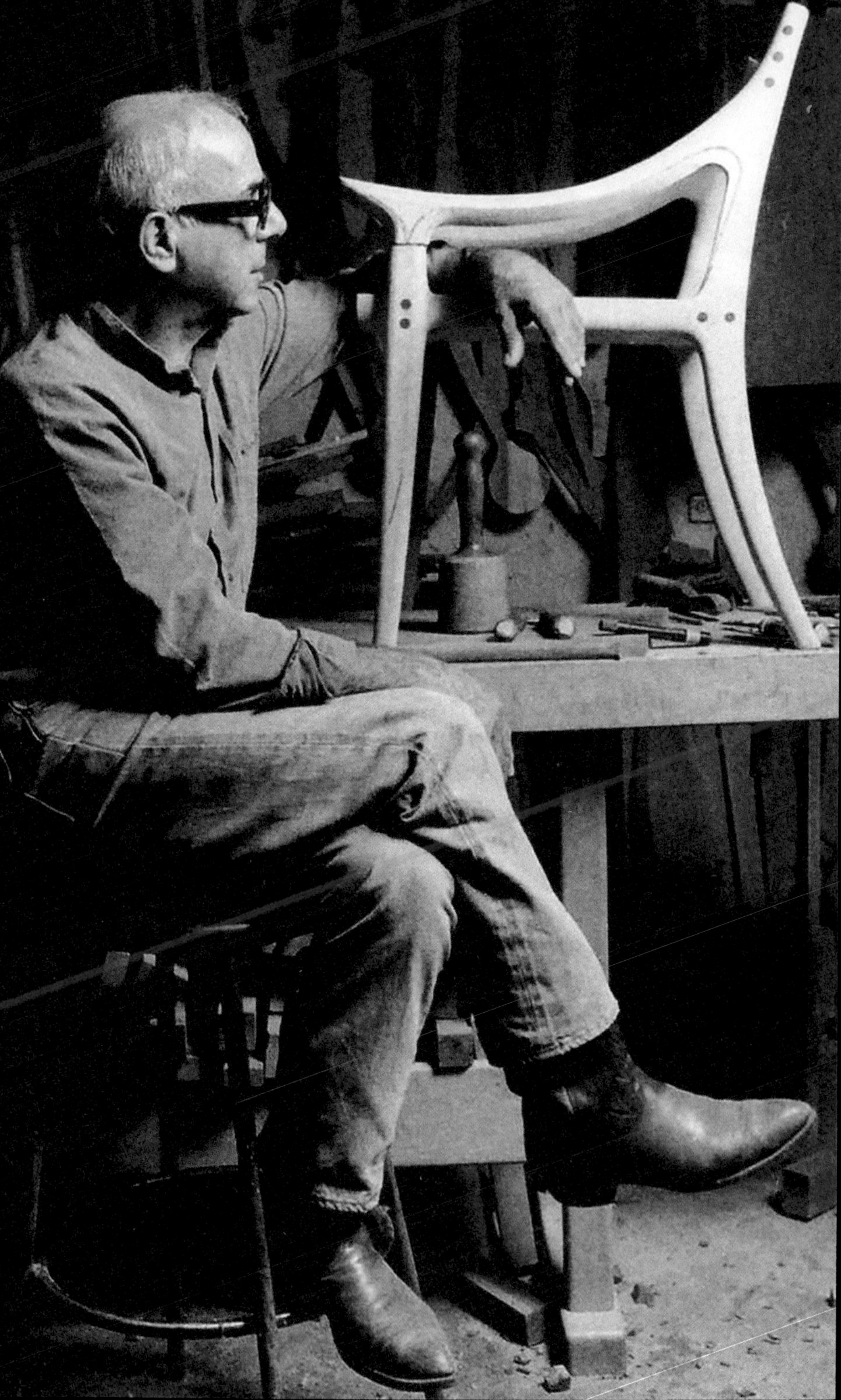

Somebody has to keep the real world running, maintained, repaired and replaced; something we realize very quickly when things don't work in our households.

Sam Maloof prepared his legacy meticulously. His philosophy and nature-related work increasingly steeped in sustainable practices will continue through his many students, trained practitioners and emulators.

He always found a way to wrap his zest in a few personal words. "I hope," he once said, "that my happiness with what I do is reflected in my furniture, that it is vibrant, alive, and friendly to the people who use it."

Sam Maloof steadfastly looked ahead. Maybe this is why, in spite of clients waiting years for delivery, he would jump to the head of the line those parents who wanted a cradle for their infants.

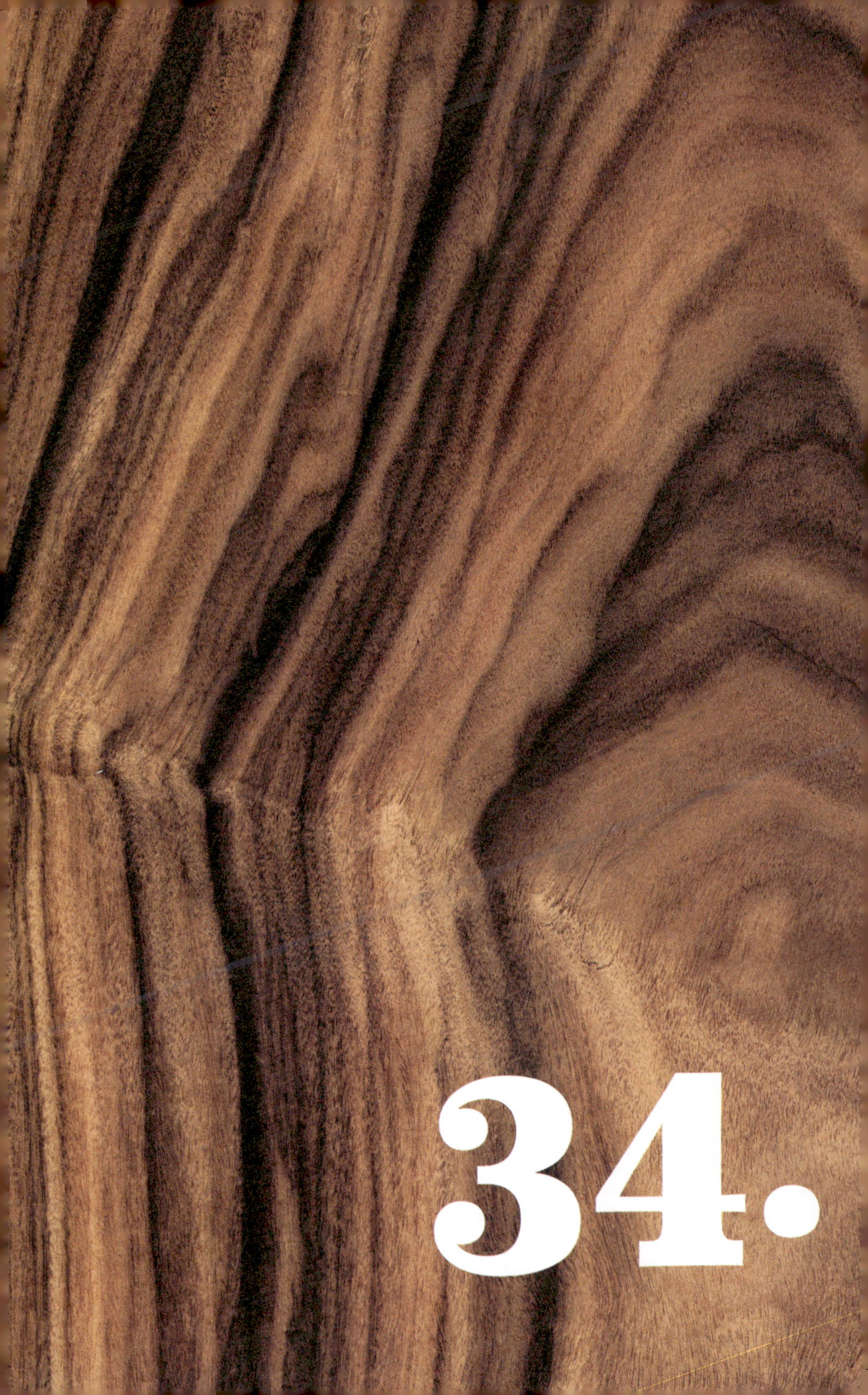

34.

CRADLE
MIKE GATTO

"My father loved Sam Maloof," said Mike Gatto. "He considered Sam not only a great artist, but a great friend."

It was a busy afternoon in Sacramento, but Assemblyman Gatto, representing the Forty-Third Assembly District in Southern California, relished the opportunity to skip lunch and squeeze between meetings and phone calls a few heartfelt words about his late father, Joe Gatto, and the man who in many ways reflected his father's personal and artistic values.

"Sam Maloof and Joe Gatto—they really were birds of the same feather. Even in height and manners. The way that they stood." Gatto leaned back in his chair and smiled at the picture in his head. "It was amazing seeing the two of them together. They had this chemistry. They almost completed each others' sentences."

Joe Gatto, jewelry artist and high school art teacher for forty-six years with the Los Angeles Unified School District, grasped in Sam's work something inspiring and instructive—something fundamentally nourishing—long before the woodworker's reputation had reached a national audience. While raising three children on a teacher's salary, plus earnings from the one or two jewelry pieces he managed to complete each school year, Joe set aside small sums that accumulated over time to purchase Sam's furniture. Starting in the early 1960s, he bought a Maloof desk, two settees, a rocking chair, and some barstools.

"We kept them immaculate," recalled Gatto. "But they were all put to use in our house. We sat in the rocker at least once a week. We sat in the settees every day. The bar stools? My father let us climb all over them. He cherished these pieces, he wanted them to be part of our family's life forever. They cost a small fortune even back then, and for a while I used to hear from my mother about what a poor investment they were." He paused, silently underscoring the miscalculation. "I think those days have passed."

Joe Gatto also purchased a piece of furniture destined to occupy the symbolic and emotional center of his family: a cradle and hutch. Over the years, it rocked to sleep all three of his children.

"My little sister is four years younger," said Gatto, "and I can picture her clear as day in the cradle. I must have been five; she was one. I own the cradle today, and both of my daughters have used it. But Sam's pieces are works of art as much as they're functional furniture. There's something that appeals to—I don't want to use the word 'heart.' I think it's in your brain stem. That's how basic the connection is. Sam's pieces are so elegant in the design, so beautiful and seamless—so otherworldly. Nobody who has seen a Sam Maloof piece isn't moved in some profound way. My father was certainly moved."

Joe Gatto's appreciation grew from his own experience as an artist, his lifelong pursuit of wresting shape and sensibility from variously resistant materials—a 14 karat gold ring fitted with an ancient glazed-blue Egyptian scarab, or a bracelet crafted from lapis and jade. Every piece, however distinct in design, accent, color, or intention, followed a scrupulous course from conception to execution. It evolved much like a Maloof chair.

"My father ate and breathed art," recalled Gatto. "He was an artist in every fiber of his being. For forty-six years, he taught school to put food on the table. People asked what he was going to do when he retired, and he said, 'I'm going to see if I can make it creating jewelry.' He was like a very talented actor who for years works as a waiter to support his family and then returns to the stage at age seventy and actually strikes it big. He delayed his gratification. And bless his heart, he acquired a following. His pieces went from selling for $400, to $1,000, to $2,000, $3,000, up to $5,000 in a period of about ten years."

Joe Gatto's creations projected an audacity of form and theme. A ring evoking *The Cost of War* was crafted from sterling silver, porcelain, doll fragments, and pink ivory wood. Tropical hardwoods, old ivory, bone, glass, stones, precious metals, shells, plastics, Japanese ojime beads, scarabs, and other relics from antiquity were collected and employed with an aim to "recycle, reuse, and restore human and natural-made materials." In an interview with *Modern Silver Magazine,* Joe noted, "Sometimes I dedicate my

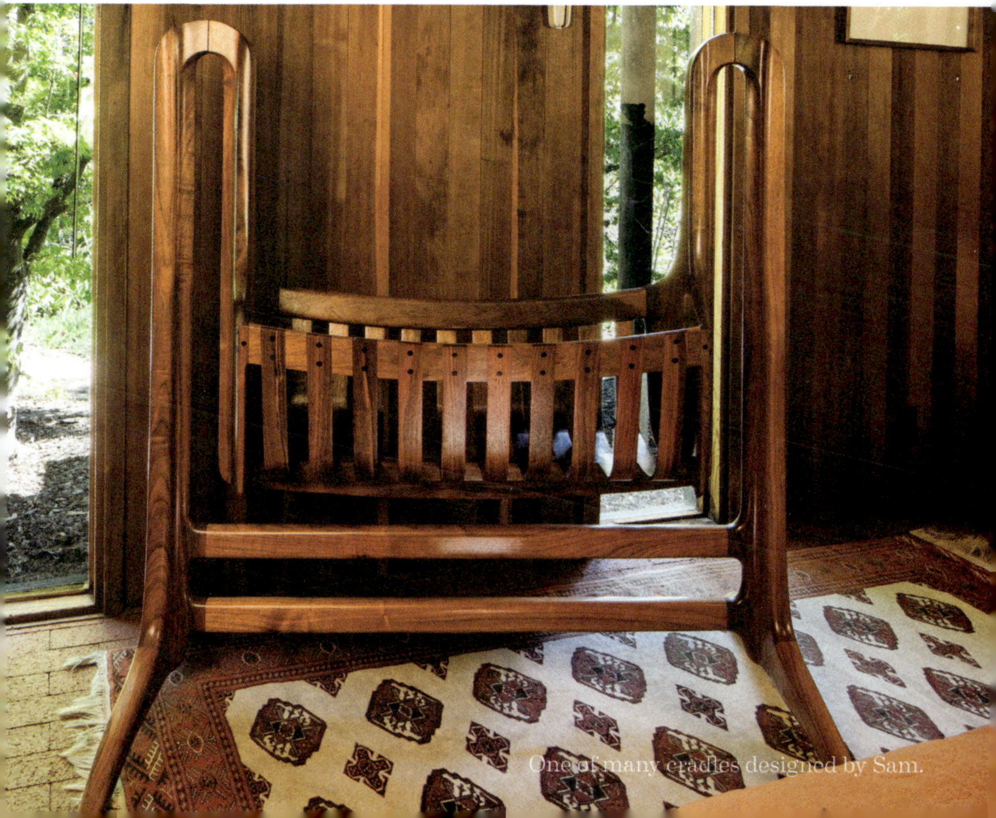

One of many cradles designed by Sam.

jewelry to my three children, Nicole, Michael, and Mariann." And in another breath: "I have dedicated all my jewelry to Earth, Sky, and Water, or Faith, Hope, and Charity."

Joe Gatto's language rang with a lofty, distinctly un-Maloofian tone. Yet the men's mutual sympathies ran deep. When the Craft and Folk Art Museum in Los Angeles reopened after an extensive remodel, Joe proclaimed that his "biggest thrill" was to have his jewelry exhibited "next to the work of another inspiring craftsman, Sam Maloof." And while Joe collected, used, and treasured an array of Maloof furniture, he relished the opportunity to reciprocate, fashioning a pair of wide band rings for Sam and Alfreda that employed small pieces of ebony provided by Sam—both artists enchanted by its grain, density, and texture.

Mike Gatto acknowledges the impact of being raised in an intensely artistic household where he could observe his father endeavoring to perfect his creations. Where he could spend his childhood climbing across Sam's furniture.

"Still," he admitted, "I don't have an artistic bone in my body. It's a little painful. You hear about families where the father was a football star and he's unhappy that son isn't following his path. In my house, it was the opposite. My father felt upset that I was so left-brained. But even those of us who don't have artistic skills can appreciate great art."

Gatto paused, overwhelmed for a moment by the profound absence of both men to their families and the world.

"There will never be another Sam Maloof," he said. "There will never be another Joe Gatto. I don't see anybody taking up that mantle. Working with your hands with such dedication. Spending the time to create something that will last. Society is poorer for losing people like that."

Yet their influence persists. In their work, certainly, but also—perhaps as importantly—by their example. Mike Gatto has been a direct beneficiary of the craftsman's code.

"I can wire a house," said Gatto. "I can take apart an engine. My father taught me all that. He operated from the premise that if there was something that needed to be done, he was going to do it with his own hands. That's something else he and Sam shared. They both had ties to Old World craftsmanship. Sam to Lebanon, my father to Southern Italy. They deeply understood a way of working that very few people in the world appreciate today. When you combine that approach to art with a good heart, as both men had, it's very, very magnetic."

The assemblyman gathers the papers on his desk, rises from his chair. It's time to return to work. There's a drought afflicting the state of California and a water bill that needs to be hammered together, revised, presented, debated, rewritten, moved through committee, and ultimately passed by both houses.

"As a legislator," explained Gatto, "you're trying to achieve a dynamic where you can serve the people well by doing what needs to be done. It takes some creativity to make sure you know what's going to appeal to the various interests, different parts of the state, disparate members of the legislature. Then you need to sell it to the people. You can't just be logical. When a politician starts talking logic and appeals only to the brain, and not to the heart, he's done."

The craft of legislation: it's a practice that not only requires rigorous thinking and skillful persistence, but also capacious vision, ambition, generosity, humility. It takes a kind of artfulness.

"I'm the only member of the legislature who doesn't accept sponsored bills," said Gatto. "With sponsored bills, somebody hires attorneys to represent their interests. They write the legislation, hand it to you, and you introduce it. I've never done a sponsored bill and I never will. If there's something I'm going to introduce, I want to write it. It's a source of pride." Gatto moved toward his office door, and then eased into the corridor. "If my name is going on the legislation, I want to have control of the process. If your name is on something, damn it, you're going to do it right!"

35.

GENERATION
STEFFANIE DOTSON

In 2006, Steffanie Dotson, then a fledgling woodworker, drove across town to San Diego's Balboa Park and watched Sam Maloof pull up outside the Mingei International Museum in his slate-gray Porsche.

"He was such a magical person," she recalled. "This charismatic rock star of a woodworker."

Steffanie had completed her bachelor's degree at San Diego State University, abandoning engineering to focus on applied design in the art department. She worked in concrete, paper, and, most influentially, wood. "I try not to lose that thirst for learning new things," she said. "That's the trick to woodworking. You can spend a life learning." Over the next two years, she sporadically enrolled in classes at Palomar College, progressively absorbing the technical skills required to balance the "concept and art side of things" that dominated academia. Palomar proved a bastion of Maloof enthusiasm. "They're huge supporters. They have a chair class and most of the students make Maloof-inspired pieces."

Sam had recently turned ninety. Not the decade conventionally associated with theatrical entrances in a flashy sports car. "Which made him more sweet," said Steffanie. "Here's somebody who started making furniture because he needed to furnish his house. And now he's selling his rocking chairs for thousands of dollars. I was like a kid admiring a basketball star—he was the Michael Jordan of woodworking for me."

Steffanie could remember her own grandfather making custom cabinets for a living. Every Thanksgiving, the scent of fresh-cut wood wafted up from the basement, mixing with the homey aroma of roast turkey and pumpkin pie. "So nostalgic for me," she sighed. Her mother's father also worked wood, but "more as a hobbyist." Steffanie watched the older men at their benches; occasionally, she was indulged and permitted to handle the tools. But even though she was "always doing crafty things—making houses out of toothpicks" or building massive forts from the huge refrigerator boxes her stepfather brought home from his job at Sears, wielding

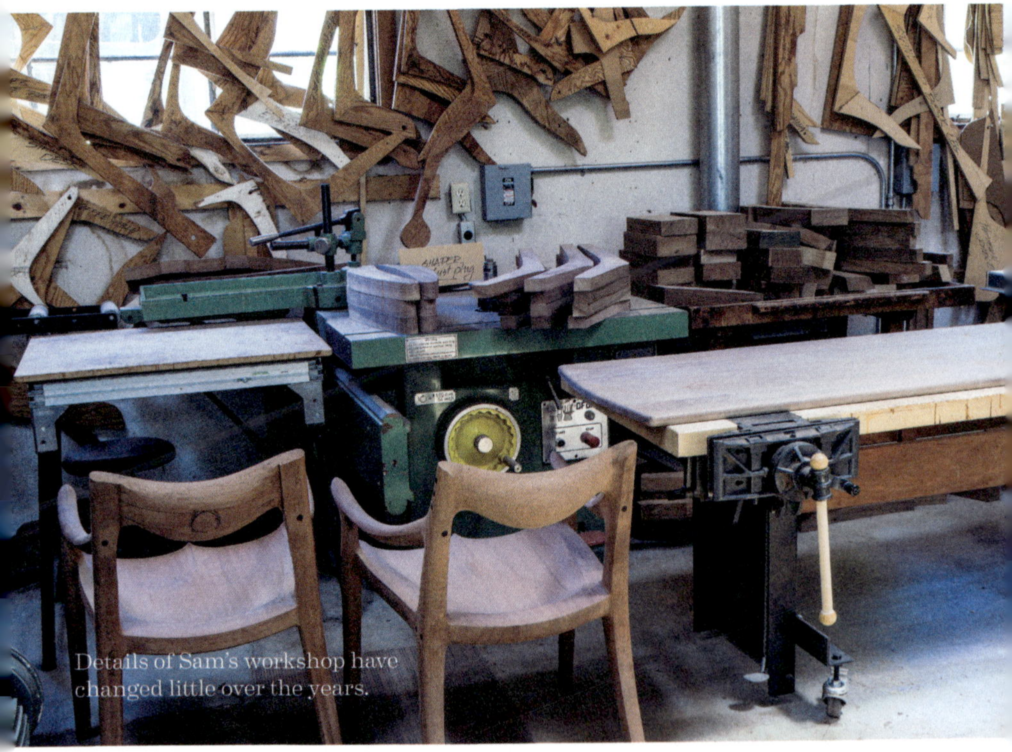

Details of Sam's workshop have changed little over the years.

a utility knife to slice open doors and windows—nobody considered teaching her about woodworking techniques. She was a girl.

"There is a curveball in there being a woman," she admitted. "I naively didn't expect to encounter these attitudes. Going to the lumber yard and being asked: 'Is this for you or your dad?'" Or the more blatant skepticism approximating rudeness. "'So, you build furniture, you have a table saw?' It's in people minds that furniture is big and heavy and you don't expect a woman to be lugging around large slabs of wood. I decided that's something I could get my feathers ruffled about or I could just move on."

She did move on, with alacrity and purpose, joining eight other woodworkers for five years in a collaborative where "critique flowed freely through the shop."

"Most people just see the end product. They don't know all the steps. Many times there are mistakes you end up fixing. That's what makes a really good furniture maker—when you figure out how to fix your mistakes. It's such a labor of love. A little bit of your soul goes into each piece."

Sam conveyed that message at the Mingei. A good woodworker invests himself—herself—in every chair, table, and cradle. It's part of the creed, the price of guild standing among bona fide artisans.

At that point in her career, Steffanie felt hungry for advice, receptive to the qualities that Sam seemed to embody. "Whether I chose to acknowledge it at the time, he had a big impact on me as I was honing my thinking about who I wanted to be as a craftsperson."

Most striking that evening: Sam's certainty about the name of his profession.

"I was new to my practice," Steffanie said. "I wasn't sure what I wanted to put on my business card. Do I want to be an artist, do I want to be a sculptor? I thought it was refreshing that Sam said, 'I'm a woodworker.' He said it was an honest word. I really liked that. Woodworking is about precision, knowing the materials, taking measurements. It does convey honesty. I'll describe myself as a woodworker, but I have no qualms about calling myself an artist or designer too. I do all those things."

Today, Steffanie is president of the Furniture Society and works out of her studio at home. Her aspirations for the future fluctuate between a determination "to get out there and make lots of furniture for people" and a continuing infatuation with "the romantic side of furniture—making great pieces, maybe turning out one or two a year and being happy with those." Whatever the direction, "I always feel fulfilled working in the studio." Recently, her life has been complicated and enriched by the birth of her first child. A baby girl whom Steffanie plans to induct into the craft of woodworking as soon as the child can safely brandish the smallest of tools.

"She's in a stage now where hands are mesmerizing to her. She's figuring out how to scoop up and grab things. I definitely want her to be involved in making some of her own toys. At a very young age, you can have amazing hand skills and dexterity."

Ideally, Steffanie will someday pass down to her daughter her own grandfather's tools. She will relate what she has learned over the course of her career, which still has decades to be elaborated and deepened. Inevitably, she insists, she will strive to convey what she learned from Sam. By his example as much as his words.

How commitment to labor can bring joy. How a woodworker should measure progress over a lifetime of minute refinements. Why it's important for every craftsperson—at every level, at any age—to bathe their efforts in discipline and love.

"Everybody should learn to use their hands at a young age," Steffanie affirmed. "It just makes you better at life."

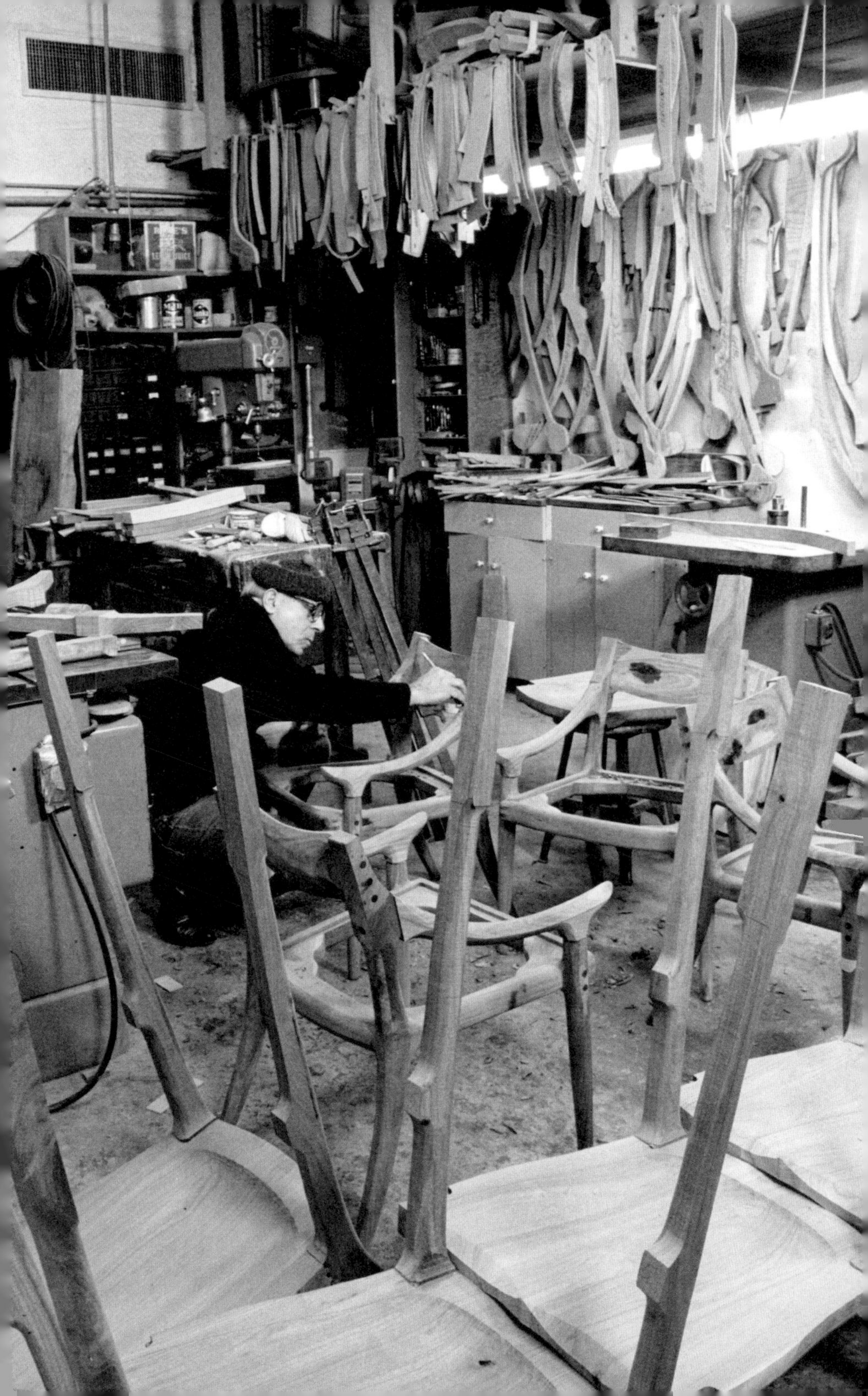

36.

LEGACY
JIM RAWITSCH

Arriving at a full stop in the garden, angled between his office, the house, and the workshop, Jim Rawitsch, executive director of the Sam and Alfreda Maloof Foundation for Arts and Crafts, smiled and spread his arms in an open-handed gesture to indicate the obvious.

"I get to work in paradise."

It's a line he relishes, and often repeats, though it's entirely sincere.

The feeling hit him the first time he visited the property—"the spirit and serenity of the place. Most people come thinking they know something about Maloof. 'Oh, he's the rocking chair guy.' Sure, the furniture is part of the story. They tell you about precision, workmanship, imagination. But the pieces become even more meaningful in context. It isn't just about a chair. It's about choices in living. If you open your eyes and heart, you can walk away from here much richer."

Recently, Jim and some board members paged through the original documents establishing the Maloof Foundation and discovered that long before the property was listed on the National Register of Historic Places, Sam and Alfreda were welcoming some three thousand visitors into their home each year on an informal basis. Over the decades, the admiring and the curious—often uninvited—wound their way to the family's front door, requesting to see the furniture and tour the grounds, hoping to meet Sam. Few left disappointed.

"Sam wasn't creating a showplace," said Jim. "He was building a home for his family and his work in a very precise and yet unlimited way. It isn't just one house—it's many houses over the years; it evolves and grows. Sam's life gets bigger and the house expands to make room for it. If we all just did that—created a place for ourselves to live creatively in relation to our families, neighbors, and work—everything else would pretty much take care of itself."

One of the challenges now facing the Maloof Foundation is finding the means to convey its story to future generations. Today, scores of people who intimately knew Sam are still living. But how long will that be true? "Memories of the man will fade," acknowledged Jim. What will remain are his creations—most significantly, his furniture and home, constituting a palpable exposition of the man's spirit and deeds. Yet the tale that the Maloof Foundation aspires to tell isn't solely about the achievements of a single artist. Rather, the grounds and garden, the house, workshop, archive, and art collection reinforce a more ambitious message about the possibilities of a creative life.

"The mythology of the lone wolf artist is something I want to refute," said Jim. "Sam's life shows us that art emerges in relation to a community. For a period of time, the Pomona Valley fueled the imaginations of a whole bunch of artists working in various media with different visions and skills. There was an exchange of ideas, energy, enthusiasm. While it's wonderful to celebrate an extraordinary person doing extraordinary things, it's also wonderful to think of the artist as just another Joe who manages to tell his story in a way that's accessible and enduring."

Although those closest to Sam might deem the proposition heresy, Jim argues that the move, despite its protracted agonies, may ultimately be judged indispensable to sustaining the Maloof legacy.

"I know it was disruptive and costly and maybe the original site was perfect the way it was. On the other hand, there's no guarantee that the house would still be standing had the freeway not gone forward. But because there was a provision to endow and protect the structure at a new location, its future is assured."

The Maloof home is part of the National Trust's Historic Artists' Homes and Studios program, one of thirty-two sites nationwide preserving properties where artists lived and worked, while providing access to scholars and the public. Sam's house has joined the residences of sculptor Augustus Saint-Gaudens, photographer Alice Austen, woodworker Wharton Esherick, and painters Frederic Church, N.C. Wyeth, Grant Wood, Winslow Homer, Georgia O'Keeffe, Lee Krasner, and Jackson Pollock.

"With many artists," explained Jim, "the opportunity to preserve their home and studio comes years and years after they're gone. Long after their personal possessions and their stories have been distributed to the winds." In the absence of committed family members, a group of enthusiasts may emerge to salvage the property, patch the leaking roof and drive out the pigeons. But the essence of the place can be lost. At the Maloof residence, Sam helped map out plans for the future during his final years, imbuing the project with the heart and character of its creator.

"Sam couldn't have known the degree to which so many people would be willing to bring fresh energy to the institution after he was gone," said Jim. "It's the reward for an immense amount of good and honest work. For having made so many friends in the process. It's a case study for how to take care of what we value as a community. It's why art is valuable. Every ounce of energy that goes in redoubles, multiplies."

And the prospects for the Maloof enterprise and legacy in coming years?

"There seems to be no limit to what's possible. The magic Sam brought to the world is here, ready to be discovered and carried forward."

MUSEUMS COLLECTING SAM MALOOF FURNITURE

Minnesota Museum of American Art, St. Paul, Minnesota 1959

Museum of Arts and Design, New York, New York 1960

Oakland Museum of California, Oakland, California 1969, 1984

The Council House, S.C. Johnson Company collection, Racine, Wisconsin 1969

Museum of Fine Arts, Boston, Massachusetts 1976

Craft and Folk Art Museum, Los Angeles, California 1980

The Vice President's Residence, The White House, Washington, DC 1980

The White House, Washington, DC 1981

St. Louis Art Museum, St. Louis, Missouri 1983

Jimmy Carter Presidential Library and Museum, Atlanta, Georgia 1985

Albuquerque Museum, Albuquerque, New Mexico 1986

Dallas Museum of Art, Dallas, Texas 1986

Los Angeles County Museum of Art, Los Angeles, California 1987

Metropolitan Museum of Art, New York, New York 1988

M.H. de Young Memorial Museum, San Francisco, California 1990

Philadelphia Museum of Art, Philadelphia, Pennsylvania 1981, 1991

Detroit Institute of Arts, Detroit, Michigan 1991

Toledo Museum of Art, Toledo, Ohio 1992

Smithsonian American Art Museum, Renwick Gallery, Washington, DC 1992, 1993

The White House Collection of American Crafts, Washington, DC 1995

SAM MALOOF (1916–2009)

1916 Born in Chino, California

1934 Graduates from high school, and works in graphic arts for Vortox

1934 Produces first furniture for family

1939–41 Employed by industrial designer Harold Graham

1941–45 Serves in US Army, stationed in the Aleutian Islands

1946–48 Returns to Pomona Valley, goes to work for artist Millard Sheets

1948 Marries Alfreda Ward

1948 First exhibition, Los Angeles County Fair

1949 Begins furniture business in Ontario, CA

1949 Son Slimen Maloof is born

1951 Furniture featured in *Better Homes and Gardens*

1951 Commissioned to design and make furniture for the Pasadena home and office of industrial designer Henry Dreyfuss

1953 Relocates to Alta Loma, CA

1954 Begins work on his first shop and residence

1957 Attends first national Conference of American Craftsmen in Asilomar, CA

1959 Minnesota Museum of American Art, St. Paul, acquires work

1960 Museum of Arts and Design in New York acquires work

1959 Works with US State Department Design Project in Iran and Lebanon

1961 Builds guest house

1962 Awarded Gold Key from National Home Furnishings League, Southern California chapter

1963 Works with US State Department Design Project in El Salvador

1967 Wins award from American Institute of Architects, Pasadena chapter, for Outstanding Excellence in Craftsmanship

1969 Becomes first recipient of the Louis Comfort Tiffany Grant for the Craftsman Apprentice Program

1969 Oakland Museum of California and the Council House in Racine, Wisconsin, acquire work

1975 Inducted into the American Craft Council College of Fellows

1976 Featured in show at the Museum of Fine Arts, Boston

1980 Vice President's house, Washington, DC, acquires furniture

1981 White House acquires furniture

1983 Autobiography published, *Sam Maloof: Woodworker*

1983 Builds spiral staircase in residence

1984 Awarded National Endowment for the Arts fellowship

1984 Declared Living Treasure by the Rancho Cucamonga City Council

1985 Awarded John D. & Catherine T. MacArthur Foundation Fellowship

1985 Jimmy Carter Presidential Library acquires work

1986 Dallas Museum of Art in Texas and the Albuquerque Museum in New Mexico acquire work

1986 *People* magazine declares him "a Hemingway in hardwood"

1987 Los Angeles County Museum of Art acquires work

1988 New York Metropolitan Museum of Art acquires work

1988 Awarded American Craft Council Gold Medal

1989 Instructional video produced, *Sam Maloof: Woodworking Profile*

1990 M.H. de Young Memorial Museum in San Francisco acquires work

1990 California State Legislature approves funds for freeway extension slated to run through the Maloof property

1991 Detroit Institute of Arts in Michigan and the Philadelphia Art Museum in Pennsylvania acquire work

1992 The Smithsonian American Art Museum, Renwick Gallery, and Toledo Museum of Art in Toledo, Ohio, acquire work

1992 Receives Honorary Doctorate of Fine Arts, Rhode Island School of Design

1994 Completes negotiations with the state and county to relocate shop and residence

1994 Establishes Sam and Alfreda Maloof Foundation for Arts and Crafts

1995 White House Collection of American Crafts acquires work

1998 Alfreda Ward Maloof dies

1999 Receives Lebanon Award for Innovation, Makhzoumi Foundation, Beirut, Lebanon

2000 Receives Most Influential People Award, *Los Angeles Times*

2001 Receives Award of Distinction, Lifetime Achievement Award, The Furniture Society

2001 Maloof residence and shop relocated

2001 Marries Beverly Wingate

2001 *The Furniture of Sam Maloof* published

2002 *The Furniture of Sam Maloof Retrospective,* Smithsonian American Art Museum, Renwick Gallery

2002 Maloof residence opens to the public

2003 Receives State of California Governor's Historic Preservation Award for relocation project

2006 *Sam Maloof 90th Birthday Retrospective,* Mingei International Museum, San Diego, CA

2007 Jacobs Education Center and Hanna Archival Conservatory completed on Maloof grounds

2007 Receives Honorary Doctorate, California State University, San Bernardino

2009 Dies at home

2010 Maloof compound listed on the National Register of Historic Places

2011 *The House That Sam Built: Sam Maloof and Art in the Pomona Valley, 1945–1985* published by The Huntington Library, Art Collections, and Botanical Gardens, San Marino, CA

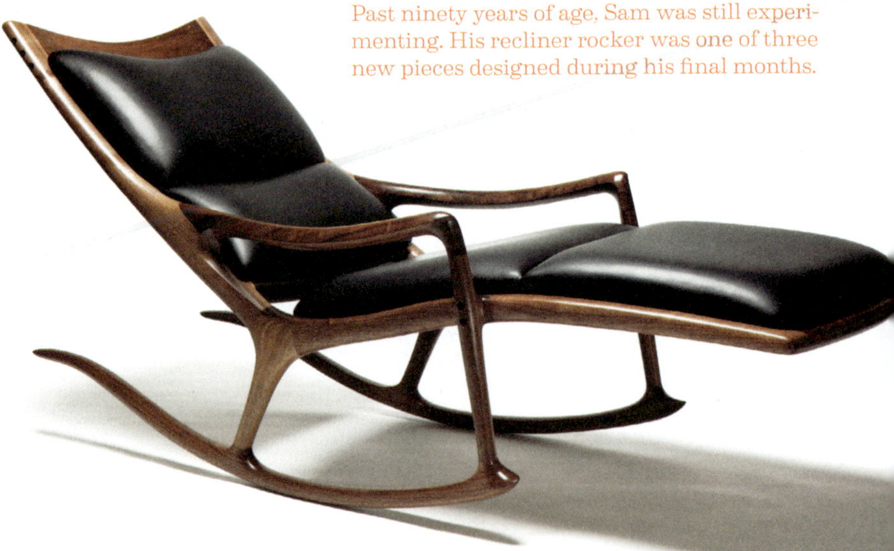

Past ninety years of age, Sam was still experimenting. His recliner rocker was one of three new pieces designed during his final months.

BIBLIOGRAPHY

Three sources proved indispensible to approaching Sam Maloof's long life and prodigious output. Jeremy Adamson's *The Furniture of Sam Maloof* is the authoritative volume, comprehensive in scope and vividly written, and blessed with Jonathan Pollock's gorgeous photographs. Sam's own account of his career, *Sam Maloof, Woodworker,* provided a helpful viewpoint with its emphasis on method and craft. Finally, *The House That Sam Built: Sam Maloof and Art in the Pomona Valley, 1945–1985,* by Harold B. Nelson, places Sam within the context of his region's lively artistic community, offering insight into the work of a captivating array of painters, sculptors, fabric artists, and woodworkers.

PUBLICATIONS

Adamson, Jeremy. *The Furniture of Sam Maloof.* Washington, DC: Smithsonian American Art Museum, 2001.

Bourne, Peter G. *Jimmy Carter: A Comprehensive Biography from Plains to Post-Presidency.* New York: Scribner, 1997.

Claremont Museum of Modern Art. *The Fiesta Artists of Padua Hills 1953–1959.* Claremont, CA: 2011.

Comba, Steve. *James Hueter: A Retrospective.* Claremont, CA: Claremont Museum of Art, 2009.

Crawford, Matthew B. *Shop Class As Soulcraft: An Inquiry into the Value of Work.* New York: Penguin Press, 2009.

Gotkin, Michael. *Artists' Handmade Houses.* New York: Abrams, 2011.

Johnson, Christy. *Harrison McIntosh: A Timeless Legacy.* Pomona, CA: The American Museum of Ceramic Art, 2009.

Kaplan, Wendy. *California Design, 1930–1965: Living in a Modern Way.* Cambridge, MA: MIT Press, 2011.

Lauria, Jo and Suzanne Baizerman. *California Design: The Legacy of West Coast Craft and Style.* San Francisco, CA: Chronicle Books, 2005.

Maloof, Alfreda Ward. *Recollections From My Time in the Indian Service 1935–1943.* Klamath River, CA: Living Gold Press, 1997.

Maloof, Sam. *Sam Maloof, Woodworker.* New York: Kodansha USA, 2013.

Mayfield, Signe S. *In the Realm of Nature: Bob Stocksdale and Kay Sekimachi.* San Diego, CA: Mingei International Museum, 2015.

McIntosh, Catherine. *The Artists of the G.I. Bill: The Impact of the G.I. Bill on the Claremont Art Community.* Claremont, CA: Claremont Museum of Modern Art, 2014.

Nelson, Harold B. *The House That Sam Built: Sam Maloof and Art in the Pomona Valley, 1945–1985.* San Marino, CA: The Huntington Library, Art Collections, and Botanical Gardens, 2011.

Owen-Towle, Carolyn Sheets. *Damn Gorgeous: A Daughter's Memoir of Millard Owen Sheets.* Oceanside, CA: Oceanside Museum of Art, 2008.

Rodel, Kevin P. and Jonathan Binzen. *Arts and Crafts Furniture: From Classic to Contemporary.* Newtown, CT: Taunton Press, 2003.

Sasse, Gene. *Maloof at 90: An American Woodworker.* Alta Loma, CA: Gene Sasse, 2011.

Starr, Kevin. *Golden Dreams: California in an Age of Abundance, 1950–1963.* New York: Oxford University Press, 2009.

———. *Material Dreams: Southern California through the 1920s.* New York: Oxford University Press, 1990.

Trapp, Kenneth R. *The Arts and Crafts Movement in California: Living the Good Life.* New York: Abbeville Press, 1993.

Watson, Sir Francis. *The History of Furniture.* New York: Crescent Books, 1976.

DIGITAL RESOURCES

Millard Sheets. Oral history. UCLA: February 5 and February 9, 1977. http://oralhistory.library.ucla.edu.

FILM & TELEVISION

The Ceramic Art of Harrison McIntosh: A Personal View of the Artist. Produced and photographed by Catherine McIntosh. Claremont, CA: McIntosh Productions, 1994.

Craft in America, volume 1. Produced by Carol Sauvion. PBS, 2007.

Design for Modern Living: Millard Sheets and the Claremont Art Community 1935–1975. Written, produced, and directed by Paul Bockhorst Productions, 2014.

ACKNOWLEDGMENTS

Sam Maloof cultivated a talent for friendship throughout his long life. In writing this book, I have relied almost entirely on his friends and family members for their recollections and insights about Sam as an artist and man. Any misinterpretation of viewpoint or error of fact is entirely my responsibility.

I especially want to thank Mike Johnson and Larry White who proved endlessly helpful and wise in detailing their long associations with Sam during their years of working in the Maloof shop.

Jim Rawitsch provided enthusiasm and guidance for this project with perfect grace and generosity.

I am grateful for Molly Woodward's skillful, painstaking, and patient editing. Ashley Ingram's design provided a perfect complement to the spirit of the project. Thanks, too, to Gayle Wattawa and Diane Lee for their oversight and guidance.

I thank Malcolm Margolin, founder and publisher of Heyday, for involving me from the start.

Thanks, one and all, for your stories and astute judgments about a complicated man and unusually fruitful artist—along with your kind assistance in so many other ways:

Tony Abeyta, Linda Apodaca, Ros Bock, President Jimmy Carter, Jennie Chinn, Hiram Chodosh, Steffanie Dotson, Larry Gandsey, Mike Gatto, Kathryn Golden, Dawn Hawk, John and Marilyn Hocking, Jim Hueter, Mike and Joanne Johnson, Gary Keene, Beverly Maloof, Ted and Gloria Maloof, Slimen Maloof, Lindell Marsh, Harrison McIntosh, Catherine McIntosh, Knox and Carlotta Mellon, Ralph Nader, Kevin Numoto, Father Peter O'Reilly, Mark Peel, Linda Roman, Carol Sauvion, Kay Sekimachi, Priscilla Sifuentes, John and Debbie Starr, Carolyn Owen-Towle and Tom Towle, Joe Unis, Larry White, and as always, Ann Van Steenberg.

ABOUT THE AUTHOR

Fred Setterberg is the author of *Lunch Bucket Paradise: A True-Life Novel* and *The Roads Taken: Travels Through America's Literary Landscapes,* winner of the AWP prize in creative nonfiction. He is the coauthor of several other books, including *Under the Dragon: California's New Culture* and *Toxic Nation,* both written with Lonny Shavelson. He is the recipient of a NEA creative writing fellowship.

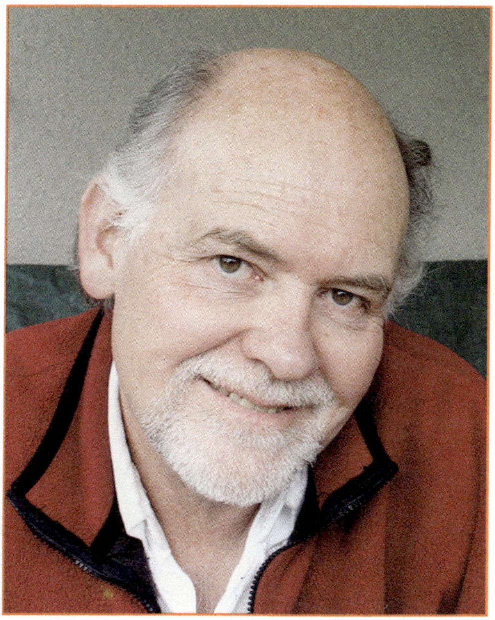

ABOUT THE SAM AND ALFREDA MALOOF FOUNDATION FOR ARTS AND CRAFTS

A not-for-profit, public benefit institution, The Sam and Alfreda Maloof Foundation for Arts and Crafts welcomes visitors for tours, exhibitions, a drought-tolerant garden, and educational programs serving youth, adults, artists, and veterans.

Centered at the historic home and workshop where celebrated woodworker Sam Maloof (1916–2009) designed and built California modernist furniture for more than half a century, The Maloof is a Smithsonian Affiliate, listed in the National Register of Historic Places, and part of the National Trust's Historic Artists' Homes and Studios program.

Located forty miles east of Los Angeles, The Maloof holds one of the nation's most extensive collections of Sam Maloof furniture works, and a unique collection of works by artists from California, the Southwest, and Mexico. The Maloof is located at 5131 Carnelian Street, Rancho Cucamonga, CA 91701. For membership, reservations, or other information, please visit www.malooffoundation.org. Phone: 909-980-0412. Email: info@malooffoundation.org.

The Maloof Foundation thanks Maloof Foundation members, docents, donors, volunteers and the Board of Directors, including (in alphabetical order): Nick Brown, Charles Field, Anthea Hartig,

Larry Henderson, Ann Joslin, Beverly Maloof, Lindell Marsh, Carolyn Owen-Towle, Richard Pace, Connie Ransom, Janice Rutherford, John Scott, David Spencer, Fritz Weis, Diane Williams, and Todd Wingate. Emeritus board members are Ted Maloof, Slimen Maloof, William Knox Mellon, and Joe Unis. We also thank Maloof staff Roslyn Bock, Linda Apodaca, Rich Deely, Melanie Swezey-Cleaves, Kristine Davis, Colleen Sullivan, Larry White, Priscilla Sifuentes, and Nidhi Gandhi.

This book was made possible with generous support from The Wingate Foundation and many others.

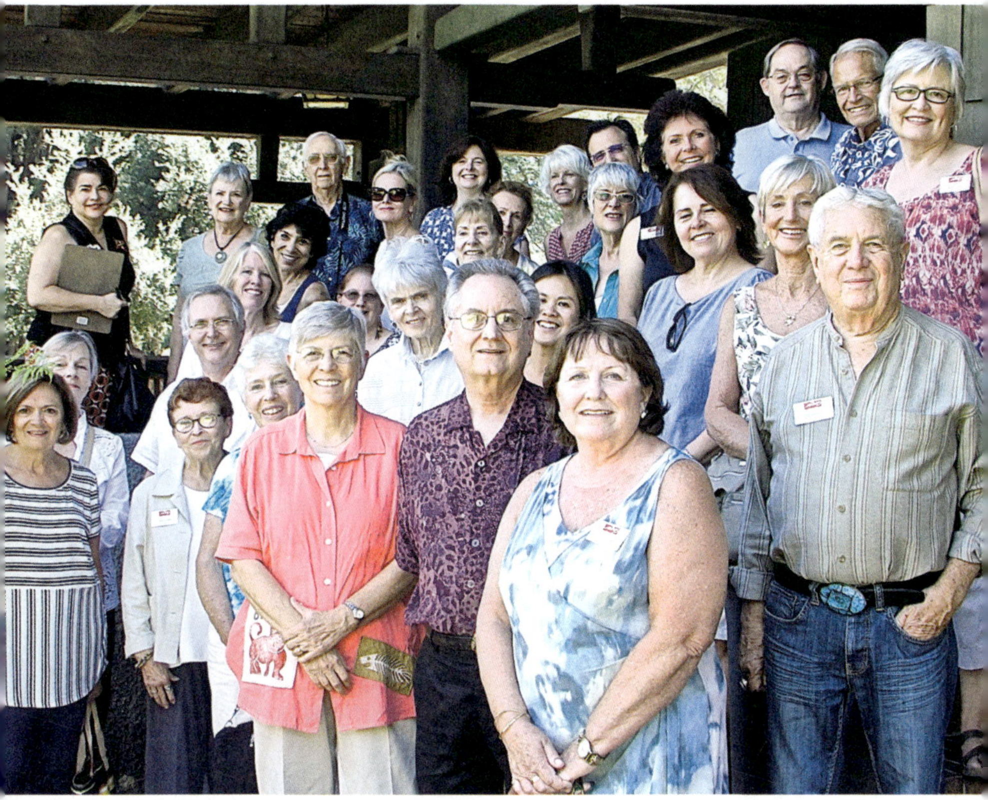

The Maloof is member supported, with docents and volunteers who carry forward the stories of Sam Maloof's legacy.

ABOUT HEYDAY

HEYDAY
into California

Heyday is an independent, nonprofit publisher and unique cultural institution. We promote widespread awareness and celebration of California's many cultures, landscapes, and boundary-breaking ideas. Through our well-crafted books, public events, and innovative outreach programs we are building a vibrant community of readers, writers, and thinkers.

THANK YOU

It takes the collective effort of many to create a thriving literary culture. We are thankful to all the thoughtful people we have the privilege to engage with. Cheers to our writers, artists, editors, storytellers, designers, printers, bookstores, critics, cultural organizations, readers, and book lovers everywhere!

We are especially grateful for the generous funding we've received for our publications and programs during the past year from foundations and hundreds of individual donors. Major supporters include:

Advocates for Indigenous California Language Survival; Anonymous (3); Arkay Foundation; Judith and Phillip Auth; Judy Avery; Carol Baird and Alan Harper; Paul Bancroft III; The Bancroft Library; Richard and Rickie Ann Baum; BayTree Fund; S. D. Bechtel, Jr. Foundation; Jean and Fred Berensmeier; Joan Berman; Barbara Boucke; Beatrice Bowles, in memory of Susan S. Lake; John Briscoe; David Brower Center; Helen Cagampang; California Historical Society; California Rice Commission; California State Parks Foundation; California Wildlife Foundation/California Oak Foundation; Joanne Campbell; The Campbell Foundation; Candelaria Fund; James and Margaret Chapin; Graham Chisholm; The Christensen Fund; Jon Christensen; Cynthia Clarke; Community Futures Collective; Lawrence Crooks; Lauren and Alan Dachs; Nik Dehejia; Topher Delaney; Chris Desser and Kirk Marckwald; Lokelani Devone; Frances Dinkelspiel and Gary Wayne; Doune Trust; The Durfee Foundation; Megan Fletcher and J.K. Dineen; Michael Eaton and Charity Kenyon; Richard and Gretchen Evans; Friends of the Roseville Library; Furthur Foundation; The Wallace Alexander Gerbode Foundation;

Patrick Golden; Erica and Barry Goode; Wanda Lee Graves and Stephen Duscha; The Walter and Elise Haas Fund; Coke and James Hallowell; Theresa Harlan and Ken Tiger; Cindy Heitzman; Carla Hills; Leanne Hinton and Gary Scott; Sandra and Charles Hobson; Nettie Hoge; Claudia Jurmain; Kalliopeia Foundation; Judith Lowry and Brad Croul; Marty and Pamela Krasney; Robert and Karen Kustel; Guy Lampard and Suzanne Badenhoop; Thomas Lockard and Alix Marduel; Thomas J. Long Foundation; Bryce Lundberg; Sam and Alfreda Maloof Foundation for Arts & Crafts; Michael McCone; Nion McEvoy and Leslie Berriman; Moore Family Foundation; Michael J. Moratto, in memory of Major J. Moratto; Stewart R. Mott Foundation; Karen and Thomas Mulvaney; Richard Nagler; National Wildlife Federation; Native Arts and Cultures Foundation; The Nature Conservancy; Nightingale Family Foundation; Steven Nightingale and Lucy Blake; Northern California Water Association; Panta Rhea Foundation; Pease Family Fund; Jean Pokorny; Jeannene Przyblyski; Steven Rasmussen and Felicia Woytak; Susan Raynes; Restore Hetch Hetchy; Robin Ridder; Spreck and Isabella Rosekrans; Alan Rosenus; The San Francisco Foundation; Toby and Sheila Schwartzburg; Stephen M. Silberstein Foundation; Ernest and June Siva, in honor of the Dorothy Ramon Learning Center; William Somerville; Carla Soracco; John and Beverly Stauffer Foundation; Radha Stern, in honor of Malcolm Margolin and Diane Lee; Liz Sutherland; Roselyne Chroman Swig; TomKat Charitable Trust; Jerry Tome and Martha Wyckoff; Thendara Foundation; Sonia Torres; Michael and Shirley Traynor; The Roger J. and Madeleine Traynor Foundation; Lisa Van Cleef and Mark Gunson; Stevens Van Strum; Patricia Wakida; Marion Weber; Sylvia Wen; John Wiley & Sons, Inc.; Peter Booth Wiley and Valerie Barth; Bobby Winston; Dean Witter Foundation; Yocha Dehe Wintun Nation; and Yosemite Conservancy.

BOARD OF DIRECTORS

GETTING INVOLVED

To learn more about our publications, events, and other ways you can participate, please visit www.heydaybooks.com.

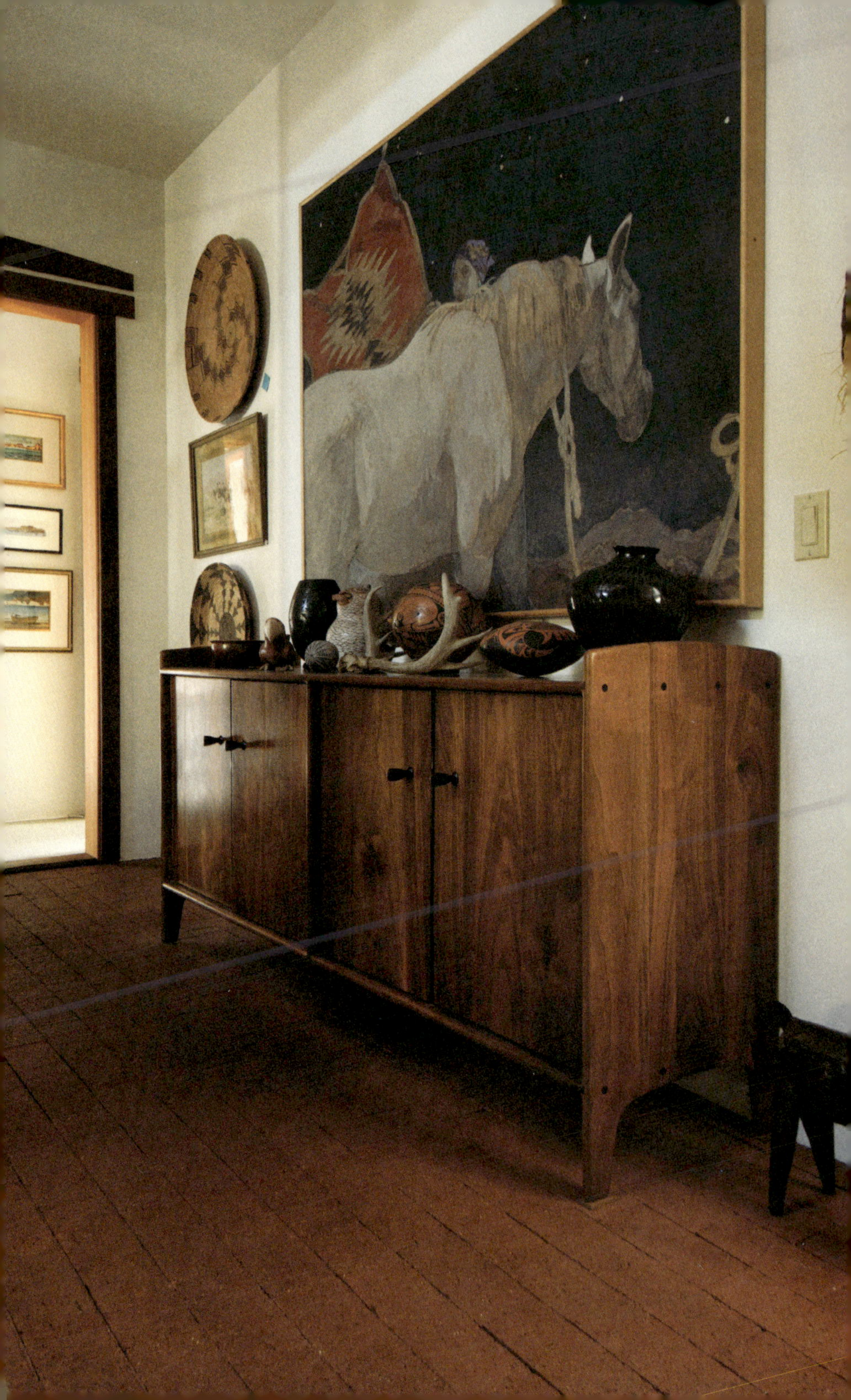